Matthew Holbeche Bloxam

**The Principles of Gothic Ecclesiastical Architecture**

Matthew Holbeche Bloxam

**The Principles of Gothic Ecclesiastical Architecture**

ISBN/EAN: 9783744644402

Printed in Europe, USA, Canada, Australia, Japan

Cover: Foto ©berggeist007 / pixelio.de

More available books at **www.hansebooks.com**

# THE PRINCIPLES
## OF
# GOTHIC ECCLESIASTICAL ARCHITECTURE,

### WITH AN

EXPLANATION OF TECHNICAL TERMS, AND A CENTENARY OF ANCIENT TERMS.

#### TOGETHER ALSO WITH

NOTICES OF THE INTERNAL ARRANGEMENT OF CHURCHES PRIOR TO, AND THE CHANGES THEREIN IN AND FROM, THE REIGN OF EDWARD VI.

*WITH NUMEROUS ILLUSTRATIONS ON WOOD,*
*Mostly by the late T. O. S. Jewitt.*

BY

MATTHEW HOLBECHE BLOXAM,

OF RUGBY, SOLICITOR.

(AUTHOR OF "A GLIMPSE AT THE MONUMENTAL ARCHITECTURE AND SCULPTURE OF GREAT BRITAIN," PUBLISHED A.D. 1834.)

*ELEVENTH EDITION.*

VOLUME II.

LONDON: GEORGE BELL AND SONS.

MDCCCLXXXII.

# CONTENTS

OF

## Vol. II.

| Chap. | | Page. |
|---|---|---|
| I. | ON THE INTERNAL ARRANGEMENT OF CHURCHES PREVIOUS TO THE REFORMATION . | 1 |
| II. | OF MONASTIC ARRANGEMENT . . | 234 |

Small Sepulchral Slab.

# LIST OF ILLUSTRATIONS

## TO

## VOL. II.

|  | Page. |
|---|---|
| ANGLO-SAXON Porch, Monkswearmouth church, Durham | 1 |
| Stoup, south door, Oakham church, Rutlandshire | 11 |
| Ancient Sacring Bell, found at Warwick | 28 |
| Seats with Carved Poppy Heads, Sunningwell church | 31 |
| Ancient Pew-work, Tysoe church, Warwickshire | 32 |
| Early English Screen, Thurcaston, Leicestershire | 36 |
| Pixis, or Ancient Stone Offertory Box, and Image Bracket, Bridlington Priory church, Yorkshire | 63 |
| Bracket, Evington church, Leicestershire | 65 |
| Norman Sculpture, Bobbing church, Kent | 66 |
| ———— Bobbing church, Kent | 67 |
| *(Kindly lent by the Council of the Royal Archæological Institute)* | |
| Brass Reading Desk, Merton College chapel, Oxford | 73 |
| Representation of 'Missa pro defunctis,' 13th century | 79 |
| Pricket Candlestick, 12th century, from a Crypt painting, Canterbury cathedral | 82 |
| Ancient Altar Candlestick of Latten, in the church of Clapton-in-Gordano | 83 |
| Thurible or Censer, 14th century | 85 |
| Chalice and Paten, Sandford, Oxfordshire, 13th or 14th century | 89 |
| Norman Piscina, Romsey church, Hants | 94 |

# LIST OF ILLUSTRATIONS.

|  | Page. |
|---|---|
| Piscina, Newnham Regis, Warwickshire | 95 |
| Ambrie or Locker | 97 |
| Low-side Window, Dallington church, Northamptonshire | 128 |
| Altar and Piscina, Warmington church, Warwickshire | 144 |
| Sculptured Stone, Yaxley church, Huntingdonshire | 149 |
| Reliquary, 14th century, Brixworth church, Northamptonshire | 152 |
| ———— Kew Stoke church, Somersetshire | 153 |
| *(Kindly lent by the Council of the Royal Archæological Institute)* | |
| Representation of a Corpse attired for Burial, Holy Trinity chapel, Stratford-on-Avon | 196 |
| Mural Painting, Dodford church, Northamptonshire | 205 |
| Fragment of Painted Glass, Canterbury cathedral | 217 |
| Pattern Glazing, Long Itchington church, Warwickshire | 219 |
| Sepulchral Arch and Effigy, 14th century, Coleshill church, Warwickshire | 233 |
| Gable Cross, Colwall church, Herefordshire | 234 |
| Sepulchral Effigy of a Cistercian Monk, Orton-on-the-Hill, Leicestershire | 285 |
| Sepulchral Effigy of a Monk, Hexham Priory church, Northumberland | 287 |
| Sepulchral Effigy of a Knight in a Friar's Cowl, Conington church, Huntingdonshire | 291 |
| Sepulchral Effigy of an Abbess, Polesworth church, Warwickshire | 297 |
| Plan of Kirkstall Abbey | 309 |
| Sepulchral Headstone Cross, Goodnestone, Kent | 311 |
| Stone Candlestick, Upton Castle chapel, Pembrokeshire | 314 |
| *(Kindly lent by the Cambrian Archæological Association)* | |
| Sepulchral Monument, Westminster Abbey, 14th century | 321 |
| Monument, Hillmorton church, Warwickshire | 324 |

Corbel Head.

Anglo-Saxon Porch, Monkswearmouth Church, Durham.

## CHAPTER I.

ON THE INTERNAL ARRANGEMENT OF CHURCHES PREVIOUS TO THE REFORMATION.

NOTWITHSTANDING the spoliation of our English churches, especially of those of conventual foundation, in the reign of Henry the Eighth, and the changes effected in the ritual and ceremonies of the Church in the reign of Edward the Sixth, and the destructive violence occasioned by the Puritans in the middle of the seventeenth century, and still more the unnecessary and wanton destruction of many vestiges of ecclesiastical and historic interest in these later times, occasioned by the so-called restoration of our churches, without any conservative feeling, our ancient churches still retain relics of the past, not as yet swept entirely away. These point to usages in religious worship, with which our ancestors were familiar, but which, some having been abrogated, and others differing in many respects from the liturgical rites of the Reformed Church, cannot be fully understood without some knowledge of the former discipline of the Church, and of the services connected with it. As historic reminiscences, however, the vestiges thus left are not without their interest and value.

Though so early as the fourth century, we meet with a variety of liturgical offices in use in the Eastern Church, which, differing from one another in minute particulars, agree in general and essential points, the formation of such liturgies is not to be ascribed to the period at which they are first found reduced into writing, but to usages of much higher antiquity, the origin of which we are perhaps hardly able to trace, though they were early considered to be coeval, or nearly so, with the apostolic age.* For the discipline and mystical rites of the Primitive Church during the three first centuries, when it

* From Justin Martyr's account of the celebration of the Lord's Supper, it is evident that there was in his time a set form of public worship. His first apology was written within half-a-century of the death of the apostle St. John.

was struggling against persecution, being performed in secret, were imparted by mere oral communication; and the most important to the faithful only, or those fully admitted into the Church, whilst the catechumens, or those converts who still remained in a state of probation, were, after the performance of a certain portion of the accustomed service, dismissed, and not allowed to remain to be partakers of the more solemn rites.

Justin Martyr adverts to the custom of the Christians turning to the east in prayer, and adds that the Church received from the apostles the mode and place of prayer.[b]

Tertullian, who wrote at the close of the second, and early in the third century, alludes to certain of the mystical ceremonies of the Christians as having been betrayed to strangers, and amongst these he mentions the custom of making the sign of the cross on the body: and Basil, Bishop of Cæsarea, circa A.D. 370, speaks of the mystical rules and discipline of the Church as originating partly from unwritten doctrine, and partly from apostolical tradition; and amongst the observances then in use, which had been traditionally handed down, and were of unknown antiquity, he particularises the trine baptismal immersion, the signing with the sign of the cross, the turning toward the east in prayer, and the use of a solemn form of words beyond those contained in the Scripture, both before and after the exhibition of

[b] In primis Justin, ad Orthodoxos respond, ad quæst. 118. ait, "ideo Christianos omnes precum tempore spectare ad Orientem" ... additque in fine, "Ecclesiam a sanctis Apostolis orandi morem et locum accepisse." *Duranti De Ritibus Ecclesiæ L. l. c. III.*

the bread of the Eucharist, and the cup of blessing; and he observes that these and many other mysteries were derived from the unwritten doctrine of a concealed and mystical tradition.[c]

The Constitutions known by the title "Apostolical," which were written about the close of the third, or early in the fourth century, contain a formulary of the Eucharistic service, as then observed in the Eastern Church. Amongst the rites referred to in this most ancient liturgy, which is also called the "Clementine," may be noticed the kiss of peace, according to the apostolical injunction, whence originated the pax of silver at a much later period presented to be kissed;[d] the ablution of hands before the offertory, originating from a Jewish rite, and the admixture of water with the wine in the cup of the Eucharist, a custom of immemorial tradition.[e] The priest also wore a white or shining garment, and in the communion the mystical elements, in both kinds, were partaken of by all the faithful.[f]

The origin of the Roman liturgy is involved in some obscurity; it has been partly ascribed to Leo and Ge-

[c] De Spiritu sancto, ad Amphilochium c. XXVII.
[d] The Pax was used at the coronation of Queen Elizabeth.
[e] The mixed cup in the Eucharist is expressly mentioned by Justin Martyr. It was enjoined in the Order of Communion, set forth in the reign of Edward the Sixth, A.D. 1547. It was likewise enjoined in the first Liturgy of the Reformed Church of England, A.D. 1549. This custom was never expressly abrogated, though in the Liturgy of 1552 and subsequent Liturgies wine only was required for the cup. I am at a loss to find the reason for the change.
[f] The ancient Greek Liturgies of the Eastern Church have within the last few years been edited and translated into English by the late Rev. J. M. Neale, D.D., and the Rev. R. F. Littledale, LL.D., rendering them more easily accessible than the more costly work of Renaudot.

lasius, both of whom flourished in the fifth century, though the latter Pontiff appears rather to have added to an old office than composed an entire new one. Gregory the Great, who flourished a century after Gelasius, revised it, and added considerably to the services and rites of the Church.[g] Much of the substance of the ancient liturgies of the East may be found comprised in that contained in the Sacramentary of Gregory; yet the order of the service is different, and the form of the latter subsequently prevailed generally throughout the West, and appears to have been introduced into Britain by Gregory himself, through Augustine. The Gregorian liturgy in order and substance, but with divers additional prayers and forms, has ever since been followed by most of the Churches in communion with that of Rome; and though the rituals of many of those Churches differ in the *ordo missœ* or ordinary of the mass, that variable part which precedes the preface, the canon of the mass, which follows it, with some additions to the post communion, continued nearly the same, word for word, as that compiled or revised by Gregory.

Prior to the arrival of Augustine towards the close of the sixth century, the ancient liturgy of the British Church is supposed to have been the same as, or derived from, that of the Gallican Church.[h] From the time of Augustine to the Reformation the liturgies of

[g] These three Liturgies are contained in a folio volume, edited by Muratori, and entitled, *Liturgia Romana Vetus Tria Sacramentaria Complectens, Leonianum scilicet, Gelasianum, Et Antiquum Gregorianum*, published at Venice, A.D. 1748.

[h] The ancient Gallican Liturgy, *De Liturgia Gallicana*, was edited and commented on by the learned Mabillon, and published at Paris, A.D. 1729.

the English Church were derived from that of Gregory, probably at first with little or no alteration. Subsequently in different districts a variety of offices prevailed. Of these the most noted was that contained in the service book known as *The Use of Sarum*, compiled by Osmund, Bishop of Salisbury, about the close of the eleventh century. This use or service was adopted throughout the greater part of England; though the cathedrals of York, Lincoln, Hereford, and Bangor had also their several uses or forms of worship, varying in some respects from that of Sarum, but the canon was the same in all.[i] It is with reference to these, and other occasional observances, that we should view the peculiar appendages which still exist in many of our ancient churches.

During the first two centuries of the Christian era we have little notice, owing perhaps to the persecutions which then prevailed, of material buildings, purposely erected and set apart for Divine worship. In the early part of the third century, however, traces appear of distinct buildings, ἐκκλησιὰ, *domus Dei*, appropriated for the purpose of Christian service, and these were not few in number, though the records we have of them are scanty. About the year 240 of the Christian era, Gregory Thaumatergus is recorded to have built a church or structure, for religious worship, of more than ordinary proportions, at Neocæsarea.

In the pseudo-Apostolical Constitutions or Canons,

---

[i] Both the ordinary and canon of the mass according to the use of Sarum, with the ceremonies used thereat before the Reformation, were translated by Fox, and appear in his Martyrology.

written probably in the latter part of the third, or early in the fourth century, we have some slight account of the plan and arrangement of these sacred edifices; "Let the building be long, with its head to the east, with its vestries on both sides, at the east end, and so it will be like a ship."[a]

At the commencement of the fourth century, in the tenth and last general persecution, by the edict of Dioclesian, that the churches should be levelled with the ground, many were destroyed.

Soon after the cessation of this persecution many churches were built, and not a few heathen temples were converted into churches.

Eusebius, in his panegyric on the building of the churches, addressed to Paulinus, Bishop of Tyre, by whose zeal principally the church of Tyre, at that time by far the most noble of the Christian structures in Phœnicia, was built, after describing generally the plan and mode of construction of that edifice, proceeds to say,—"For when he (the builder) had thus completed the temple, he also adorned it with lofty thrones, in honour of those who preside, and also with seats decently arranged in order throughout the whole, and at last placed the holy altar in the middle; and that this again might be inaccessible to the multitude, he inclosed it with framed lattice work, accurately wrought with ingenious sculpture, presenting an admirable sight to the beholders."

'Εφ' ἅπασί τε τὸ τῶν ἁγίων ἅγιον θυσιαστήριον ἐν μέσῳ θεὶς αὖθις καὶ τάδε ὡς ἂν εἴη τοῖς πολλοῖς ἄβατα, τοῖς ἀπὸ ξύλου

[a] Whiston's Translation.

περιέφραττε δικτύοις εἰς ἄκρον ἐντέχνον λειτουργίας ἐξησκημένοις ὡς θαυμάσιον τοῖς ὁρῶσι παρέχειν τὴν θέαν.

Thus early in the fourth century the distinction between the different portions of a church, the body and sanctuary, which we now designate as nave, and chancel, or choir, is shewn to have existed.

Few and brief are the notices by St. Chrysostom and Gildas, of the religious structures used for worship by the ancient British Church. The altars therein only are mentioned, and of such structures we have now no visible traces. From the time of Augustine to the Norman conquest, the Anglo-Saxon churches appear from existing remains to have been small compared with the Norman churches. Some consisted of a nave and chancel only, as at Wittering and Bradford-on-Avon; many had a tower westward of the nave; some had aisles, as at Brixworth and Repton; some were built in the form of a cross, with transepts, as at Worth, and Stanton Lacey, and Stow, Lincolnshire; some terminated with a semicircular apse, as the original chancel at Brixworth; some with a polygonal apse, as at Wing. Most, however, appear to have been rectangular at the east end. Some had the tower between the chancel and nave, as at Wootton-Wawen; some had crypts or subterraneous passages, as at Repton, Hexham, Ripon, and Wing; a few had projecting porches, as at Bradford-on-Avon. All had the chancel pointing eastward.

Four churches in the north of England, Hexham, Ripon, Jarrow, and Monkswearmouth, all erected in the seventh century, all noticed and described by ancient writers, still retain vestiges of their original construction. The two former were built by Archbishop Wilfrid, the two latter by Benedict Biscopus. Richard, Prior of Hexham, circa A.D. 1180, describes the building of that church, A.D. 674. The substructure consisted of crypts and subterraneous oratories and winding passages. The apse was adorned with histories and images, and figures sculptured in' relief on the stone, and coloured paintings. In the oratories, both within and beneath the church, altars were constructed in honour of the blessed Virgin, of St. Michael, of St. John the Baptist, of the holy Apostles and others; relics of saints, books, vestments and utensils of the church were numerous. Such another church could not at the time of its erection be found on this side the Alps.[1] Some of the winding passages of the church built by Wilfrid are still accessible inside of the present church. These, constructed of materials from some old Roman building, have been only partially cleared out. The monastic church of Ripon was likewise built by Wilfrid. Some vaulted passages and small chambers are all the vestiges of the original church. Benedict Biscopus, A.D. 676, built the church of the monastery of Monkswearmouth, having sent for masons from Gaul, to construct it of stone after the Roman manner. He also sent to Gaul for glass factors, to glaze the windows; and he decorated it with paintings of the blessed Virgin and the twelve Apostles, with subjects taken from the Evangelists, and the visions of the Apocalypse, with these the walls were covered. He obtained also from abroad sacred vessels and vestments, and from Rome

[1] Rich. Pr. Hag. int. X Scriptores.

a multitude of books and relics."' The tower of this church and the porch still remains, together with some rude sculptured ornaments. The church of the monastery of Jarrow, likewise erected by Benedict Biscopus, was completed and dedicated A.D. 685. This church the founder decorated with paintings representing corresponding events in the Old and New Testament, such as Isaac carrying the wood on which, being bound, he was placed, and our Lord bearing his cross. The brazen serpent uplifted in the wilderness, and the Son of Man affixed to the cross."

The ALTARS in the Anglo-Saxon churches, as we see by illuminated MSS. had each an altar covering, and a cross standing upon them. Of the altar furniture we have an account of that presented by King Ina, A.D. 708, to one of the chapels of the abbey church, Glastonbury, amongst the costly church plate and articles, composing which, were a chalice, paten, and thurible of gold; candlesticks and a vessel for holy water of silver, images of gold and silver of our Lord, the blessed Virgin, and the twelve Apostles, and altar coverings and sacerdotal ornaments, wrought with gold and precious stones.

On entering a church through the porch on the north or south side, or at the west end, we sometimes perceive on the right hand side of the door, at a convenient height from the ground, often beneath or within a canopied niche, or *fenestella*, and partly projecting from the wall, a stone basin: this was the *stoup*, or receptacle for holy water, called also the *aspersorium*, into

---

"' Vita S. Benedicti, auctore Ven. Bede.
" Vita S. Benedicti, Bede.

which each individual dipped his finger, and crossed himself when passing the threshold of the sacred edifice. The custom of aspersion at the church door appears to have been derived from an ancient usage of washing, as an emblem of purity, before prayer.* The

Stoup, south door, Oakham Church, Rutlandshire.

stoup is sometimes found inside the church, close by the door, but the stone basin appears to have been by no means general, and probably in most cases a move-

* The stoup was a vessel, says Durantus, which held in churches the holy water with which those were accustomed to be sprinkled who entered. And he subsequently observes, "*Institutum fuit vasa ista aquæ benedictæ ad ostium ecclesiæ a latere ingredientis, ubi potest dextro collocari. In ceteri testamento non nisi lotus templum ingrediebatur. Cœterum vas istud aquæ benedictæ e marmore lapideve solido, non lateritio, nec spongioso fieri debet, adspergillum que decens e labro catenula appensum habere. Durantus, de labro, seu vase aquæ benedictæ.*" Zosimus also speaks of this

able vessel of metal was provided for the purpose; and in an inventory of ancient church goods at St. Dunstan's, Canterbury, taken A.D. 1500, we find mentioned "a stope off lede for the holy wat$^r$ atte the church dore." We do not often find the stone stoup of so ancient a date as the twelfth century; one much mutilated, but apparently of that era, may however be met with inside the little Norman church of Beaudesert, Warwickshire, near to the south door. At Ecton church, Northamptonshire, inside the north porch on the right, is a stoup under an ogee-headed canopy, trefoiled within. On each side of the west entrance of Irthlingborough church, Northamptonshire, is a stoup. On the right hand side of the entrance into the tower of Cerne Abbas church, Dorsetshire, is a stoup. On the right side of the west door of Mutchelney church, Somersetshire, is a fine stoup, consisting of an angular-shaped basin, with quatrefoiled compartments on the sides: this is supported by an angular-shaped shaft panelled, and converging ribs, the fan-like compartments between being also panelled. The basin is beneath a plain canopy. At Rowington, church, and Ilmington church, Warwickshire; Stanton Harcourt church, Oxfordshire; Fordington church, Dorsetshire; St. Benedict's church, Glas-

custom, "*Erat autem Romanis vetusta consuetudo, ut quum, limen templi transeundum esset, sacerdos secundum morem ethnicum madidos quosdam olivæ ramusculos manu tenens ingredientes aspergebot.*"—*Zosim. l. vi. c. vi.* —Of the origin of this custom the learned Benedictine Martene in his work *De Antiquis Monachorum ritibus*, published A.D. 1690, thus treats, "*Circa medium sæculum nonum præceptum de benedicendi diebus Dominicis aqua reperio in quadam homilia Leonis Papa IV. de cura pastorali in qua sic lego: omni die Dominica ante missam, aquam benedicite, unde populus aspergatur et ad hoc proprium vas habete.*" p. 138.

tonbury; Sedgeberrow church, Gloucestershire; and at many other churches are stoups, some within a fenestella or niche, others simply projecting from the wall. The author of *The Rites of Durham*, edited by Davies, of Kidwelly, A.D. 1672, in his description of that church as it was before the Reformation, says:—"There were two fair holy water stones belonging to the Abbey church of Durham, of a very fair blew marble. The fairest of them stood within the north church door over against the said door, being wrought in the corner of the pillar, having a very fair shrine of wainscot over head very finely painted, with blew and little gilt stars, being kept very clean, and always fresh water provided against every Sunday morning, wherein one of the monks did hallow the said water very early in the morning before divine service. The other stood within the south church door."

Amongst the *Articles of Visitation*, by Bishop Bonner, A.D. 1554, one is, " Whether there be at the entry of the church, or within the door of the same, an holy water stock or pot, having in it holy water to sprinkle upon the enterer, to put him in remembrance both of his promise made at the time of his baptism, and of the shedding and sprinkling of Christ's blood upon the cross for his redemption; and also to put him in remembrance, that as he washeth his body, so he should not forget to wash and cleanse his soul, and make it fair with godly and virtuous good living, and finally to put him in remembrance, that as water passeth and slideth away, so he shall not tarry and abide in this world, but pass and slide away as the water doth."

The PORCH was often of a considerable size, and had frequently a groined vaulting, with an apartment above, the latter being sometimes an after addition. It was anciently used for a variety of religious rites, for before the Reformation considerable portions of the marriage[p] and baptismal services, and also much of that relating to the churching of women, were here performed, being commenced "*ante ostium ecclesiæ*," and concluded in the church; also in the "*ordo ad faciendum sponsalia, sive matrimonium; In primis statuantur vir et mulier ante ostium ecclesiæ*," &c.; and these are set forth in the rubric of the Manual or service-book, according to the use of Sarum, containing these and other occasional offices.[q] We have but few porches of the Anglo-Saxon era; that at Monkswearmouth church, and that on the north side of the nave of the little Anglo-Saxon church at Bradford-on-Avon, may be enumerated as such. We have Norman porches with rooms over, generally of a later addition, and we have porches of a later era, with rooms over, erected at the same time. There is reason for believing these rooms were occupied by anchorites or recluses, of whom and of their habitations within churches, there is much to be said. In some districts we find in the porch just above the doorway into the church a small narrow loft or gallery, access to which

---

[p] "At the grand southern entrance of Norwich Cathedral are the espousals or sacraments of marriage, carved in stone."—*Bloomfield, vol. IV. p.* 22.

[q] In the year 1604 there was published at Douay a book, which is in reality the Sarum Manual, "*Sacra Institutio Baptizandi, matrimonium celebrandi*, &c., juxta usum insignis Ecclesiæ Sarisburiensis." Of the book, which is exceedingly rare, I have a copy.

was obtained by a very narrow staircase, constructed in the thickness of the wall of the porch. Authority as to the precise use of this gallery has not been obtained. It is however conjectured that it was for the purpose of decking the image of the patron saint, which stood in a niche over the doorway, on festival days. The south porch of Caldicot church, Monmouthshire, is large, and an erection of the fifteenth century; within it and over the door leading into the church has been a small loft or gallery, access to which was obtained by a flight of steps in the east wall of the porch. In the niche over the doorway and above the loft is a mutilated effigy of the blessed Virgin. In the south porch of the church of Weston-in-Gordano, Somersetshire, a narrow flight of steps on the east side leads up to the remains of a small gallery extending across the porch on the north side; of this gallery the moulded beams of the floor only remain, these cut across the apex of the inner doorway; in the wall above this gallery is a canopied niche, with a bracket at the foot for an image. This gallery is nine feet nine inches in length, and only two feet ten inches in width. In the south porch of Portishead church, Somersetshire, the same arrangement exists; the porch, a structure of the fifteenth century, is lofty, and the remains of the gallery which extend across it on the north side, are approached by steps within a projection on each side; above the gallery in the wall over the inner door is a niche. In the porch of Clapton-in-Gordano, in the same county, is a projection on the north side, with a doorway now stopped up. In the wall above the inner door is a niche. A niche also occurs over the inner doorway of the porch, which is large, at Kingston Seymour church, in the same county. On each side of the porch of Wraxhall church, in the same county, is a projection containing a staircase, which appears to have led to a small gallery over the inner door, of which now no vestiges remain; over the inner doorway is a niche for an image. Besides this arrangement there is a chamber over the porch, access to which is from a staircase leading from the south aisle of the church. These are the only churches I have met with where this singular arrangement exists; with the exception of Caldicot church, they are all situate in one particular district, and within a few miles of each other.

The position of the porch was in general one bay easternmost of the north or south-western portion of the church, leaving one window in the north or south wall westward. In many of the churches in Norfolk a provincialism prevailed of having the porch placed at the westernmost bay, as at Westwick church, Trunch church, Filby church, Martham church, Hingham church, Deopham church, Rickenhall church, Morley St. Botulph's church, Tunstead church, Wigenhall church, and South Creak church, Norfolk; of Southfleet church, Kent; of Freslingfield church, Suffolk; and of Merstham church, Surrey. But this practice is not to be commended for imitation.

Having entered the church, the FONT is generally discovered towards the west end of the nave, or north or south aisle, and near the principal door; such, at least, was in most cases its original and appropriate position: this was for the convenience of the sacramental rite there administered; part of the baptismal service (that of making the infant a catechumen) having been performed in the porch or outside the door,' he was introduced by the priest into the church, with the invitation, *Ingredere in templum Dei, ut habeas vitam æternam et vivas in sæculorum;* and after certain other rites and prayers the infant was carried to the font and immersed therein thrice by the priest, in the names of the three Persons of the Holy Trinity. By an ancient ecclesiastical constitution a font of stone or other durable material, with a fitting cover, was required to be placed in every church in which baptisms could be administered;' and it was, as Lyndwood informs us, to be capacious enough for total immersion. I have met with no font of an earlier period than the twelfth century. Some ancient fonts are of lead, as that in Dorchester church, Oxfordshire, and that in Childrey church, Berkshire; both of these are cylindrical in shape, and of the Norman era, encircled with figures in relief; those on the font at Dorchester representing the twelve Apostles, whilst those on that of Childrey are of bishops. Leaden fonts are also to be met with in the churches of Brookland, Kent; Wareham, Dorsetshire; Warborough, Oxfordshire; and Walmsford, Northamptonshire. Square and cylindrical or truncated cone-like shaped fonts, of Norman design, supported on a basement by one or more shafts, and either plain or sculptured, are numerous; we sometimes find on them figures of the twelve

r "Ad valvas ecclesiæ."—*Ordo ad Faciendum Catechumenum, Manuale.*
• Constitutions of Edmund, Archbishop of Canterbury, A.D. 1236. De Baptismo et eius Effectu.

Apostles, sculptured in low relief; the baptism of our Saviour also was no uncommon representation. Those bearing the Evangelistic symbols were also not uncommon, as at Fakenham, Norfolk; St. John Sepulchre, Norwich; and Stoke-by-Nayland, Suffolk. Fonts subsequent to the Norman era are not so frequently covered with sculptured figures, though such sometimes occur; they are cup-like, sexagonal, septagonal, or octagonal in shape, but the latter greatly predominate; and the different styles are easily ascertained by the architectural decorations, mouldings, tracery, panel-work, and sculpture, with which they are more or less covered. They are generally cased inside, or lined with lead, with a perforation at the bottom of the basin to let off the water. On the sides of rich fonts of the fifteenth century representations of the seven sacraments, according to the Church of Rome,[1] were not unfrequently sculptured, as on that in Farningham church, Kent. Some fonts bear inscriptions, as that of Bridekirk, Cumberland, which is Runic. The Norman font of Little Billing church, Northamptonshire, of a plain jar-like form, is likewise inscribed. The font of Keysoe church, Bedfordshire, bears an inscription in Norman French round the lower part of the basin of an octagonal decorated font; at Bradley church, Lincolnshire, is inscribed,— Pater noster ave maria and crede leren pe childe pt is nede. In the panels round the basin of the font, which is of the fifteenth century, of Bourne church, Lincolnshire, is an inscription which occupies seven sides, the eighth

---

[1] *Viz.*, Baptism, Confirmation, Eucharist, Penance, Orders, Matrimony, and Extreme Unction.

being placed against a pier; it is as follows,—𝔍𝔥𝔠 𝔢𝔰𝔱 𝔫𝔬𝔪 𝔮𝔟𝔢 𝔰𝔲𝔭 𝔬𝔪𝔢 𝔫𝔬𝔪. Round the upper part of the octagonal basin of the rich font, at St. Mary, Beverley, Yorkshire, is inscribed,—𝔓𝔯𝔞𝔶 𝔣𝔬𝔯 𝔱𝔥𝔢 𝔰𝔬𝔲𝔩𝔢𝔰 𝔬𝔣 𝔚𝔩𝔩𝔪 𝔉𝔢𝔯𝔶𝔣𝔣𝔞𝔯𝔢 𝔇𝔯𝔞𝔭𝔢𝔯 𝔞𝔫𝔟 𝔥𝔦𝔰 𝔴𝔶𝔟𝔦𝔰 𝔴𝔥𝔦𝔠𝔥 𝔪𝔞𝔟𝔢 𝔱𝔥𝔦𝔰 𝔉𝔬𝔫𝔱 𝔬𝔣 𝔥𝔦𝔰 𝔭𝔭𝔢𝔯 𝔠𝔬𝔰𝔱𝔢𝔰 𝔱𝔥𝔢 𝔵 𝔡𝔞𝔶 𝔬𝔣 𝔐𝔞𝔯𝔠𝔥𝔢 𝔶𝔢 𝔶𝔢𝔯𝔢 𝔬𝔣 𝔬𝔲𝔯 𝔏𝔬𝔯𝔡 𝔐𝔅𝔛𝔛𝔛. The fonts at Lullington, Somersetshire, and at Stanton Fitzwarren, Wilts, a very rare and emblematical sculptured font of late Norman work, are also inscribed. We find many plain sided octagonal shaped fonts of the fourteenth and fifteenth centuries, evidently intended for future decoration. The sides of the square Norman font at Thornbury church, Gloucestershire, have been subsequently decorated with Early English foliage and sculpture. We have no fonts, which from their details, we can clearly ascribe to the Anglo-Saxon era.* The covers to some rich fonts, especially to some of those of the fifteenth century, were very splendid, in shape somewhat resembling that of a spire, but the sides were covered with tabernacle-work, and decorated at

---

* It is much to be regretted that of late years many ancient fonts have been cast out of our churches, and earthenware and pewter basins substituted in their stead for the administration of the holy sacrament of baptism: a practice not authorised by the Anglican Church, but rather condemned; for in the canons set forth by authority, A.D. 1571, it is provided that "Curabunt (Œditui) ut in singulis ecclesiis sit sacer fons, *non velvis*, in quo baptismus ministretur, isque ut decenter et munde conservetur." And in the canons of 1603, after alluding to the foregoing constitution, and observing that it was too much neglected in many places, it is appointed "That there shall be a font of stone in every church and chapel where baptism is to be ministered; the same to be set in the *ancient usual places*." In the orders and directions given by Bishop Wren, A.D. 1636, to be observed in his diocese of Norwich, we find it enjoined, "That the font at baptism be filled with clear water, and no dishes, pails, or basins be used in it or instead of it."

the angles with small buttresses and crockets. Fonts with rich covers of this description are to be found in the churches of Ewelme, Oxfordshire; of North Walsham, and of Worstead, Norfolk; and of Sudbury, and of Ufford, Suffolk; and at Fosdyke church, Lincolnshire. Plain font covers of a spire-like form, and sometimes crocketted, are not uncommon: such occur at Combe Bissett, near Salisbury, and Ashby St. Ledgers, Northamptonshire. Near the font, Wraxhall church, Somersetshire, affixed to a pier, is a small stone desk for the manual or service book, in which the office pertaining to the sacrament of Baptism was contained. In Yolgrave church, Derbyshire, is a font, apparently of the thirteenth century, of a plain cup-like fashion, resting on a plain cylindrical shaft; projecting from the basin is a smaller basin, for the purpose, apparently, of containing the chrism or oil used in baptism. In Trunch church, Norfolk, the font, octagonal in shape, and of the fifteenth century, is placed beneath a curious canopy, altogether independent of the font, supported by six standards of carved work. The canopy exhibits indications of painting, amongst which are St. Mary and St. John, and is surmounted by a boss. It is of the latter part of the fifteenth, or early part of the sixteenth, century. There are other painted font covers in Norfolk worthy of attention.

The general situation of the TOWER or campanile is at the west end of the nave; it is sometimes, however, found in a different position, as at the west end of a side aisle, which is the case with respect to the churches of Monk's Kirby and Withybrooke, Warwickshire; or on one side of the church, as at Eynesbury church, Huntingdonshire, and Alderbury church, Salop; and the tower of the latter church is covered with what is called the saddle-back roof, having two gables—a peculiarity to be found in some few other churches, as the church towers of Tynwell, Rutlandshire; Bradwell, Bedfordshire; Maidford and Cold Higham, Northamptonshire; Begbrook, Oxfordshire; and others. In cross churches the tower was generally, though not always, erected at the intersection of the transept, and between the nave and chancel. Some churches have their tower or campanile completely detached, as at Berkeley, Gloucestershire; Elstow, Bedfordshire; Marston Morteyne, Bedfordshire; Woburn, Bedfordshire; Terrington St. Clements, near King's Lynn; Walton, Norfolk; Chittlehampton, Devon; Westbury-on-Severn, Gloucestershire; Mylor, near Falmouth; Ledbury, Herefordshire; Llangyfelach, Glamorganshire; Henllan, Denbighshire; Fleet, Lincolnshire; and Flixborough, Lincolnshire. Many of these towers will perhaps appear on examination, when placed on marshy soil, as in the north of Norfolk, to have been so disposed apart from the church, lest the settlement, if joined to the church, should dislocate the main structure. Towers were also occasionally used, up to the fourteenth century, as parochial fortresses, to which, in time of sudden and unforeseen danger, the inhabitants of the parish resorted for a while. The tower of Rugby church, Warwickshire, a very singular structure, apparently built in the fourteenth century, appears to have been erected for this purpose; it is of a square form, very lofty, and plain in construction, and is without a single buttress to support it; the lower windows are very narrow, and at a considerable distance from the ground—some of them are, in fact, mere loop-holes; the belfry windows are *square-headed*, of two lights, simply trefoiled in the heads, and divided by a plain mullion; the only entrance was through the church; it has also a fire-place, the funnel for the conveyance of smoke being carried up through the thickness of the wall to a perforated battlement; and it altogether seems well calculated to resist a sudden attack. Other church towers of early date appear to have been erected for a double purpose: that of a campanile, as well as to afford temporary security. The towers of Newton Arlosh church, of the church of Burgh-on-the-Sands, and of Great Salkeld church, Cumberland, appear to have been constructed with a view to afford protection to the inhabitants of those villages upon any sudden invasion from the borders of Scotland, and for that purpose were strongly fortified.[z] Sometimes they were used as beacons. On the top of the tower of

---

[z] The 28th decree of a foreign council, that of Wirtzburgh, held A.D. 1278, prohibits the fortifying of churches in order to make use of them as castles.

In 1483 Bishop Dudley granted license to John Kelyng, then Rector of Houghton-le-Spring, county of Durham, "To enclose, fortify, and embattle a tower above the lower porch, within his manse." The practice of embattling Parsonage houses of the higher class, was, for an obvious reason, not uncommon in the north. The Rector of Redmershall had a similar license in 1462; and in a list of Northumbrian fortresses taken during the minority of Henry VI., the tower of Whilton, belonging to the Rector of Rothbury, and six other fortified parsonages, viz., *Turris de Corbrigg, Vicarii ejusdem; Turris de Stanfordham, Vicarii ejusdem; Turris de Chatton, Vicarii ejusdem; Turris de Ellysden, Rectoris ejusdem; Turris de Pontiland, Vicarii ejusdem; Turris de Emylden, Vicarii ejusdem;* are enumerated amongst the *Fortalicia* or lowest order of Castelets. All these were

Hadley church, Middlesex, is affixed an iron cresset fire pan, or pitch pot. This was used and fired so late as 1745. Some church towers, especially in the counties of Norfolk and Suffolk, are round and batter, or gradually decrease in diameter as they rise upwards; most of these are of the Norman, though some are in the Early English, style; that at Little Saxḫam church, Suffolk, may be adduced as a specimen. These would appear to be so constructed, the material being flint, from the absence of local ashlar stone to work in at the angles. Spires in some instances appear to have served as landmarks, to guide travellers through woody districts, and over barren downs. The spire of Astley church, Warwickshire, now destroyed, was so conspicuous an object at a distance, that it was denominated the lantern of Arden. The spires of the churches of Monk's Kirby and Clifton, in the same county, now also destroyed, were formerly noticed as eminent landmarks.

In the towers or campaniles attached to or forming part of a church, the principal church bells were hung. Although bells may be traced to a period much earlier than the commencement of the Christian era, St. Paulinus, Bishop of Nola in Campania, in the early part of the fifth century, is said to have introduced them into churches—hence the terms Nola,[y] Campana and Campa-

probably mere Towers added, as at Houghton, to the main building, or sometimes, as at Rothbury, placed a little distant, for the advantage of situation, and calculated merely to preserve a few valuables from a sudden surprise of the borderers. The church towers were used for the same defensive purposes, and some reliques of old armour are still frequently preserved in the northern churches.—*Surtees' Durham*, vol. i. p. 157.

[y] Nola est signum seu campana, a Nola civitate in qua primus campanarum usus inventus est.—*Martene de Ant. Mon. ritibus.*

nile, the one a bell, the other a bell turret or tower. That bells were used by the Anglo-Saxons before and in the days of Bede is clear from the account he gives of the vision of Bega on the death of St. Hilda:—*Hæc tunc in dormitorio sororum pausans audivit subito in ære notum campanæ sonum quo ad orationes excitari vel convocari solebant cum quis de sæculo fuisset evocatus.*[z] They are also mentioned in the excerpts of Ecgbert, Archbishop of York, A.D. 750 :—*ut omnes sacerdotes horis competentibus diei et noctis suarum sonent ecclesiarum signa, et sacra tunc Deo celebrent officia, et populos erudiant quomodo aut quibus Deus adorandus est.* In the " Canones editi sub Edgaro rege," A.D. 960, allusion is made to them, *Docemus etiam ut justo stato tempore campana pulsetur.* And again in the constitutions of Archbishop Lanfranc, A.D. 1072, *Pulsentur omnia signa.—Sonantibus ad missam signis.—Majora signa pulsentur.—Sonet secretarius minimum signum.* Amongst the constitutions of the cathedral church of Lichfield, A.D. 1194, is one ' *De modo pulsationem.*' In this is mentioned, ' *campanam minimam,*' ' *campanam dulcem,*' ' *grossarum campanarum.*'[a] Classicum appears to have been a term applied to omnium campanarum pulsatio.

Amongst the illuminations to the MS. Benedictional of St. Aethelwold, executed in the latter part of the tenth century, between A.D. 963—984, one contains the representation of a campanile, with bells hanging in it.

In the Middle Ages, from the fourteenth to the middle of the sixteenth century, bell-founding as an art was

[z] *Bede L. iv. c. xxiii.*
[a] The above excerpts are taken from *Wilkins' Concilia, vol. i.*

PREVIOUS TO THE REFORMATION. 25

in request, and bell-founders left their marks or cognizances behind them, and sometimes their names. Of their devices we have numerous varieties. On few bells, however, previous to the middle of the sixteenth century, are there any dates. There are exceptions. One is preserved at St. Chad's church, Claughton, Lancashire, with the date 1296, as follows:— 

+ ANNO DNI· M· CC· NON AI· 

the v being reversed; at Cold Ashby, Northamptonshire, is a bell bearing the date 1317, as follows:—

+ MARIA : VOCOR : ANO : DNI : M° CCC° : XVII.

Two bells with the date 1323 are preserved in the tower of St. Mary's church, Somercotes, Lincolnshire. And at Loddington church, Northamptonshire, is, or was, a bell bearing the following inscription:—

"Mille quadringentis octogintaque duobus annis fusa fui lapsis ab origine Christi a genetrise piu protervis dicta Maria."

We have not now in our churches many bells of a period anterior to the fourteenth century; of those of a period subsequent to the middle of the sixteenth century, which form the majority of the bells in our churches, notice will be made in a subsequent chapter.

Of inscribed bells of the former period some appear with invocations or in honour of Saints. A few instances will suffice. At Sysonby, Leicestershire, is one inscribed, '*In honore sancti Leonardi.*' At Glen Magna, in the same county, one is inscribed, '*Iste campana facta est in honore Sti Cuthberti.*' At Clopton, Northamptonshire, '*Inves Joye fecit me in honore Sci Petri.*' Sometimes the name of the founder appears, as on a bell at All Saints, Leicester,—

'*Jhohannes De Strafford fecit me in honore De Marie.*'
This John Stafford is conjectured to have been a bell-founder at Leicester sometime during the fourteenth century. A bell-founder of this name was living at Leicester in 1371.

At Brentingby, Leicestershire, one of the bells bears the salutation, '*Ave Marie Gratia Plena.*' This was probably not uncommon. Bells bearing invocations to Saints, as at Aston Flamville, Leicestershire, '*Sca Caterina ora pro nobis,*' and '*Sce Leonarde ora pro nobis*'; at Willoughby Waterless, Leicestershire, '*Sancte Lavenci ora pro nobis*'; and at Wood Newton, Northamptonshire, '*Sancta Margarita ora pro nobis.*' On a bell at Fawesly, Northamptonshire, '*Sancte Botolfe ora pro nobis*'; at Clopton, in the same county, '*Sancte Petre ora pro nobis*'; and at Thrapston, in the same county, '*Sancta Anna ora pro nobis.*'

Outside the roof of some churches, on the apex of the eastern gable of the nave is a small open arch or bell cot, in which a single small bell was suspended. This was the Sanctus or Sacring Bell, thus placed that being near the high altar it might be more readily rung, when in concluding the Ordinary of the Mass, the priest pronounced the *Ter sanctus*, to draw attention to that more solemn office, the Canon of the Mass, which he was now about to commence; it was also rung at a subsequent part of the service, on the levation of the host and chalice after consecration for adoration.[b] This arch re-

---

[b] "In elevatione atque utrius que Squilla pulsatur."—*Durandi Rationale, lib. iv.*

mains on the gable of the nave of many churches, but the bell thus formerly suspended is retained in few, amongst which may be mentioned those of Long Compton, Whichford, and Brailes, in Warwickshire; and Weston-in-Gordano, Portbury, and Portishead, Somersetshire, where this bell is still preserved hung in an arch at the apex of the nave, with the rope hanging down between the chancel and the nave.[c] Mention of this bell is thus made in the Survey of the Priory of Sandwell, in the county of Stafford, taken at the time of the Suppression: " Itm the belframe standyng between the chauncell and the church, w$\underline{t}$ a litle *sanct$^m$* bell in the same." In the little church of Gumfreston, near Tenby, the ancient Sanctus bell is preserved, lying loose; it is eight inches high, and appears to have been originally suspended. In 1870 an ancient Sanctus bell, about five inches high, was found in a putlog hole in the western wall of the south aisle of Bottesford church, Leicestershire. This also appears to have been suspended. In chantry chapels, and at the altars in side aisles, a small hand bell, *tintinabulum*, was carried and rung at proper times in the service by the Acolyte; and in inventories of ancient church furniture we find it often noticed as " *a sacringe* bell;" but in an inventory of goods belonging to the chapel of Thorp, Northamptonshire, it is described as " a little *Sanctus bell*." A small sacringe bell, of bell metal, with the exception of the clapper, which was of iron, was in 1819 discovered on the removal of some rubbish from the ruins of St. Margaret's Priory,

[c] In Yeovil church accounts A.D. 1457, is an item, "*In una cordul empt p le salsyngbelle* ij$^d$."—*Collectanea Topographica, vol. iii. p.* 130.

Barnstaple. Some years ago a small sacring bell was found on the site of a religious house at Warwick.[d] On the

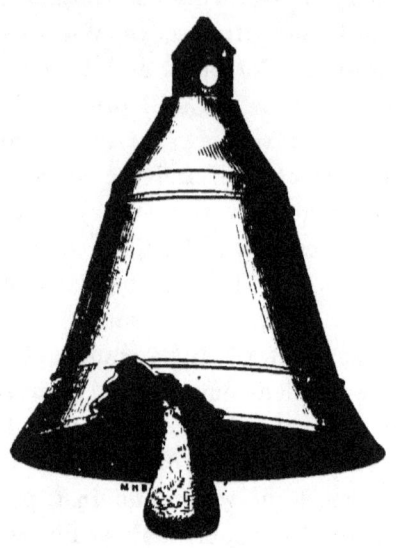

Ancient Sacring Bell, found at Warwick.

reconstruction, some years ago, of the church of Church Lawford, Warwickshire, a small sacringe bell was found.

The Benediction of Bells, a somewhat lengthy service, is of considerable antiquity, circa A.D. 970; Pope John XIII. is said, with a certain ceremony to have blessed a new bell, belonging to the church of St. John, Lateran, and this is asserted to be the first instance recorded of such benedictions. Of church bells cast subsequently

---

[d] When I made my drawing for the above vignette, this bell formed part of the celebrated Warwickshire collection of the late William Staunton, Esq., of Longbridge House, Warwick, destroyed in the disastrous fire at the Birmingham Library.

PREVIOUS TO THE REFORMATION.    29

to the middle of the sixteenth century notices will appear in a future chapter.*

Anciently the body of the church appears to have contained no other fixed SEATS for the congregation than solid masses of masonry, raised against the walls, and forming long stone benches or seats. Benches of this description, fourteen inches high and twelve inches wide, run along the north, west, and south walls of the little Norman church of Parranforth, Cornwall; a structure only twenty-five feet long, and twelve feet six inches in width. In the Norman conventual church, Romsey, Hants, plain stone benches of this description occur; they are likewise to be met with in Salisbury, and other cathedrals; and in not a few of our ancient parish churches. Against the south wall of the south aisle of Kiddington church, Oxfordshire, is a stone

---

* Campanology has of late years become a favourite and interesting study, and, ere long, the church bells in most of our counties will have received their due share of attention.

The bells of Cambridgeshire have been or will be treated upon by Dr. Raven.
Those of Cornwall by Mr. Dunkin.
Those of Devon by Mr. Ellacombe.
Those of Derbyshire by Mr. Jewitt.
Those of Gloucestershire by Mr. Ellacombe.
Those of Leicestershire by Mr. North.
Those of Norfolk by Mr. L'Estrange.
Those of Northamptonshire by Mr. North.
Those of Rutland by Mr. North.
Those of Somersetshire by Mr. Ellacombe.
Those of Sussex by Mr. Tyssen.
Those of Yorkshire by Mr. Bolter and Mr. Hope.

The above list has been furnished me by my friend Mr. H. J. Elsee, now in the Sixth Form at Rugby School, to whom I am further indebted for information in preparing this my brief notice of ancient church bells. Mr. Elsee bids fair to become an expert campanologist. To *The Church Bells of Leicestershire*, edited by my friend Mr. North, I have also to acknowledge my sincere obligations.

bench. Tickenham church, Somersetshire, has a stone bench which abuts against the west wall of the south aisle. Portbury church, in the same county, contains stone benches ranging against the south and west walls of the south aisle, and north and west walls of the north aisle. Ufford church, Northamptonshire, contains stone benches, ranging against the walls of the aisles. Haddon church, and Water Newton church, Huntingdonshire; and Yarwell church, Northamptonshire; also contain stone benches or seats for the congregation. Seats for the use of the congregation are noticed in the synod of Exeter, held A.D. 1287.*f* Open wooden benches or pew-work are rarely, if at all, met with of an earlier era than the fourteenth century, and even during that era the examples left are comparatively few. The church of Dunton Bassett, Leicestershire, retains some open bench seats, which from the crest moulding on the top of the backs indicate them to be of the fourteenth century. At Finedon, in Northamptonshire, the body of the church and aisles are almost entirely filled with low open seats with carved tracery at the ends, disposed in four distinct rows, so that the whole of the congregation might sit facing the east. These appear to be of the latter part of the fourteenth or early in the fifteenth century. The church of Byfield, in the same county, and the little interesting

*f* Item audivimus quod propter sedilia rixantur multoties parochiani, duobus vel pluribus unum sedile vendicantibus; propter quod grave scandalum in ecclesia generatur, et divinum sæpius impeditur officium; statuimus quod nullus de cætero quasi proprium sedile in ecclesia valeat vendicare, nobilibus personis et ecclesiarum patronis duntanat exceptis si qui orandi causa primo ecclesiam introlerit juxta propriae voluntatis arbitrium sibi eligat oraudi locum.

church of Shotswell, Warwickshire, both contain their original fittings of pew-work, of the early part of the fifteenth if not of the fourteenth century. To the same period we may ascribe those capital plain specimens of ancient pewing with which the beautiful church of Hawton, Lincolnshire, is filled. But it was in the fifteenth century that the practice of pewing the body of the church with low open seats generally prevailed. We find examples in this era, varying from extreme plainness to excessive richness. A bold bead moulding, one inch and a half in diameter, roving along the top of the back of the seats and returning at the ends, which were frequently panelled, has an excellent effect. Sometimes the ends or standards of the seats were carried up above the back and finished with richly

Seats with carved Poppy Heads, Sunningwell Church.

carved poppy heads. In some of the churches in

Norfolk and Suffolk rich work of this description abounds. This is also the case in Somersetshire, where rich wood-work seats are often carved in block, or in the solid. Trull church, near Taunton, is full of elaborately carved wood-work of the middle of the sixteenth century; the ends of the seats are richly carved in the solid. On one seat end is the figure of a man clad in a short coat or tunic, with trunk hose, stockings and shoes. There is also the linen pattern panel, over one of which this is carved: "*John Waye Clarke here Simon Warm am maker of thys work Anno Dni* 1560." In the latter part of the fifteenth, and early part of the sixteenth century, the panels of pews or seats were often carved with what is called the linen pattern, common in wood-work of this period. Plain seats sometimes return with elbows, as at Tysoe church, Warwickshire.[s]

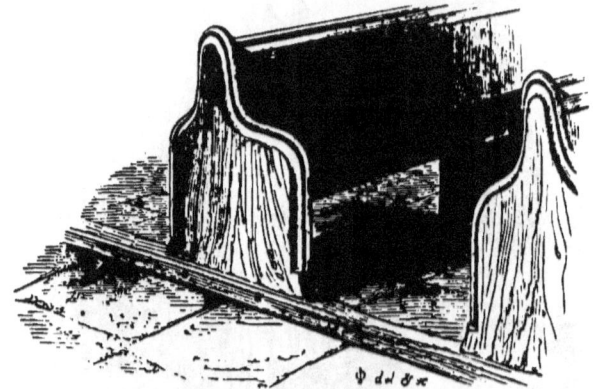

Ancient Pew Work, Tysoe Church, Warwickshire.

The PULPIT was anciently disposed towards the eastern

[s] Testamentary bequests, for the pewing of churches, were not unfrequent: thus, Wm. Bruges, Garter King at Armes, London, by his will,

part of the body of the church, but not in the centre of the aisle. Pulpits are now rarely to be found of an earlier date than the fifteenth century, when they appear to have been introduced into many churches, though not to have become a general appendage. Ancient pulpits of that era, whether of wood or stone, are covered with panel-work tracery and mouldings; and some exhibit signs of having been once elaborately painted and gilt. Mention, however, is made of pulpits at a much earlier period; for in the year 1187 one was set up in the abbey church, Bury St. Edmunds, from which, we are told, the abbot was accustomed to preach to the people in the vulgar tongue and provincial dialect.[k] The most ancient pulpit, perhaps, existing in this country, is that in the refectory of the abbey (now in ruins) of Beaulieu, Hampshire: it is of stone,

dated in 1449, gave certain monies to be bestowed upon "the complesshyng and ending of the church at Staunford, that is in coveryng with lede, glassyng and making of pleyn desques, and of a pleyn rode lofte, and *in puying* of the seyd church nourt curiously but pleynly and in paving of the hole chirch Body and Quere with broad Holand Tyle."—*Test. Vetus.* John Yonge of Herne, by will dated in 1456, gave to the fabrick of the church of Herne, viz., to make seats called *puying* X marks. Amongst the documents relating to the church of Bodmin, Cornwall, is a contract, dated in the seventh year of Henry VII. A.D. 1491, for making chairs, seats, and a pulpit for that church, "that the sayde Matthy More, carpynter, shall make or do to be made yn the parysh Churge of Seynt Petrok yn Bodmyn fully newe chayrs & seges, and iiij ranges thurgh oute all the body of the said Churge after the furme & makyng of the chayrs & seges yn Seynt Mary churge of Plympton . . . . . . and a conveoyent pulpyte yn the sayde Prysh Churge of Bodmyn after the furme & makyng of the pulpyte yn the parysh churge of Mourton yn hemstede." The consideration for the above work was to be £92.—*The Bodmin Register, p.* 33.

[k] Anglice sermocinari solebat (Abbas Samson) populo, sed secundum Linguam Norfolchie . . . . unde et pulpitum jussit fieri in ecclesia et ad utilitatem audiencium et ad decorem ecclesie.—*Cronica Jocelini de Brakelonda, sub anno* 1187.

and partly projects from the wall, and is ornamented with mouldings, sculptured foliage, and a series of blank trefoiled pointed arches, in the style of the thirteenth century. In the refectory of the ancient conventual buildings, attached to the cathedral church at Chester, is another ancient stone pulpit, apparently of the thirteenth century. The ancient refectory pulpit of the ancient abbey at Shrewsbury is apparently of the fourteenth century. There is not, however, that I am aware of, any pulpit in our churches of earlier date that the fifteenth century. The church of the Holy Trinity, at Coventry, contains a fine specimen of a stone pulpit of the fifteenth century. In Rowington church, in the county of Warwick, is a stone pulpit of the same age as that at Coventry, but much plainer in design. At Long Sutton church, Somersetshire, is a splendid wooden pulpit of the fifteenth century, painted and gilt; and the sides are covered with ogee-headed niches, with angular-shaped buttresses between. In the Devonshire churches are many ancient pulpits, both of stone and wood; some of the latter have been painted and gilt. In Norfolk some of the panels of the ancient wooden pulpits still retain paintings of the four Doctors of the church:—St. Augustine, St. Ambrose, St. Gregory, and St. Jerome; but the pulpits of this era may be distinguished without difficulty by the peculiar architectural designs they exhibit. The wooden pulpit in Lutterworth church, Leicestershire, so often pointed out as one in which Wycliffe preached, is of a date much later than his time, and apparently of the latter part of the fifteenth century. In Trull church, Somersetshire, is a richly carved pulpit with Gothic canopies at the sides, beneath which are representations of a Lily pot, of a Bishop, of a Cardinal in his Hat, and of St. Mary Magdalen; this appears to be of the middle of the sixteenth century.

We now approach the division between the nave or body of the church, and the chancel or choir. This, as we have seen, was an arrangement as early as the fourth century. I am not aware that there are now existing in this country any churches, or remains of churches, of higher antiquity than the latter half of the seventh century. The width of the chancel arch in our Anglo-Saxon and Early Norman churches is generally exceedingly narrow[i], and there does not appear to have been any screen or division of wood, but the chancel was separated from the body of the church by a curtain or veil extending across the chancel arch. Such veil indeed is mentioned in an ancient Anglo-Saxon Pontifical, "*extenso velo inter eos et populum;*"[k] and such is alluded to by Durandus, who wrote in the thirteenth century (he died A.D. 1296), "*interponatur velum aut murus inter clerum et populum.*"

The earliest wooden screen-work I know of in this country is that of St. Nicholas, at Compton, in Surrey. Here the chancel is low and groined, and of Norman architecture of the twelfth century, and above is a loft,

[i] The widths of several early chancel arches are given in Vol. I. p. 59.

[k] Durandus in his description of a church makes no mention of screen-work, but observes, "*Notandum est quod triplex genus veli suspenditur in ecclesia videlicet quod sacra operit quod sanctuarium a clero dividit, et quod clerum a populo secernit,*" evidently alluding in the latter to the curtain extended across the chancel arch.

opening into the church westward; and it is across the western boundary of this loft, formed by the Norman chancel arch of considerable width, but of no great height, that this screen, consisting of a series of semi-circular arches springing from cylindrical shafts, with moulded bases and caps, is placed.

Almost the only wooden screen of the thirteenth century, and the earliest chancel screen I have met with in this country, is one now removed, it is to be regretted,

Early English Screen, Thurcaston, Leicestershire.

from its original position in Thurcaston church, Leicestershire—Thurcaston being the birthplace of Bishop Latimer. This consists of plain panel-work in the lower part, and of a series of open arches above, trefoiled in the heads, and springing from slender cylindrical shafts, with moulded bases and caps, but not annulated. In Staunton Harold church, Oxfordshire, is a screen of a somewhat later period—of the close of the thirteenth, or early part of the fourteenth, century; the lower parts of this is composed of plain and close panel-work, the upper part of open trefoiled-headed arches, springing from slender round annulated shafts, with moulded bases and caps.

Specimens of screen-work of the fourteenth century, though not very numerous, are more common than those of an earlier period. Such, or remains of such, occur in Beaudesert church, Shotswell church, Long Itchington church, and Wolfhampcote church, Warwickshire; St. John's church, Winchester; Cropredy church, Oxfordshire; Norfleet, Kent; Geddington, Northamptonshire; and elsewhere. These are distinguished from the screen-work of the fifteenth century by the light round annulated shafts which support the open arches or flowing tracery of open work above the plain panel-work in the lower division of the screen.

In Leeds church, Kent, the open screen-work resembles a series of pointed arched windows filled with mullions and tracery. This kind of screen-work is also to be met with in Somersetshire.

Chancel screens of the fifteenth and early part of the sixteenth century are so numerous that I do not think it necessary to particularize examples. They exist from comparatively plain to enriched and elaborate carved work; the uprights are moulded, and support a horizontal cornice, richly carved with vine leaves and grapes;

whilst in the lower division of the screen the close panels are sunk, foliated in the heads, and are often painted with figures of Saints, bearing their peculiar symbols. Many of these are still visible on the screens of churches, especially of those in Norfolk. The open work in the upper division of these screens is composed of carved perpendicular tracery, supported and divided by moulded uprights, and finished with a horizontal crest moulding. Stone screens, both of the decorated and subsequent style, are occasionally met with. The rood-loft generally projected in front of both sides of the screen, so as to form a kind of groined cove, the ribs of which sprang or diverged from the principal uprights of the screen beneath, and this cove supported the flooring of the loft.

Where no rood-loft existed, as was the case in some small churches, the crest of the chancel screen served to support the rood or image of the crucifix, with the attendant images of St. Mary and St. John.

An earlier date than the eleventh century can hardly be assigned for the introduction of the ROOD with the figures of St. Mary and St. John into our churches, though in illuminated manuscripts, somewhat before that period we find such figures pourtrayed with a crucifix.[1] In the Abbey church, Bury St. Edmunds, the rood and the figures of St. Mary and St. John, which were placed over the high altar, were, (as we are informed by Joceline, who wrote his Chronicle in the twelfth century,) the gift of Archbishop Stigand."

---

[1] *Cottonian MS. Titus D. xxvii.* 10th sæc.

" " Crux que erat super magnum altare, et Mariola et Johannes, quas imagines Stigandus archiepiscopus magno pondere auri et argenti ornaverat, et sancto Ædmundo dederat."—*Cronica Jocelini de Brakelonda*, p. 4.

Gervase, in describing the work of Lanfranc in Canterbury Cathedral, as it appeared before the fire, A.D. 1174, speaks of a pulpit or loft with a transverse beam across the church, which separated the choir from the nave, and which sustained a great cross with two cherubims and the images of St. Mary and St. John. *Pulpitum vero turrem* (he is speaking of the central tower) *predictam a navi quodammodo separabat. Supra pulpitum trabes erat per transversum ecclesiæ posita quæ crucem grandem et duo cherubim et imagines sanctæ Mariæ et sancti Johannis Apostoli sustentabat.*

We find occasional mention of the rood-loft in ancient documents, as in wills. Thus, William Bruges, Garter King-of-Armes, at London, by his will bearing date the 26th February, 1449, gave certain monies to be bestowed upon "the complesshying and ending of the church of Staunford," and amongst other things for the making of "a pleyn rode lofte."

John Fane of Tunbridge, by his will dated April 6th, 1488, bequeathed to the high altar of the church of Tunbridge xxs., and to the structure of the rood loft thereof x marks, on condition that the churchwardens build it within two years.

Joan Viscountess Lisle, widow, by will dated 8th August, 1500, after directing her body to be buried in the parish church of St. Michael upon Cornhill, adds,— "Also I will that my executors cause to be made and set up on the high rood loft in the said church of St. Michael, two escotcheons; the one of them with the arms of my right noble lord and husband, the Viscount Lisle, and my own arms jointly; and the other of the arms of my right worshipful husband, Robert Drope, and my own jointly; to the intent that our souls by reason thereof, may the rather be there remembered and prayed for."

Richard Starkey of Stretton, in the county of Chester, by will dated 29th May, 1526, gave as follows:—"Itm. I bequeth towards the making to the roode seler at Budworth vi*s*. viij*d*.".

Matthew Beke, by will dated 22nd November, 1520, "I bequeth unto the roode seller of Manchester, when y$^t$ shall be *p* cede xl*s*."

William Walton, Priest, by will dated 7th January, 1527, bequeathed his body to be buried in the church of Croston, "undr the rodeceller afore the chauncell."

Thurstan Tyldisley, of Wardley, by will dated 1st September, 1st Edward VI., "I bequeth towards ye byulding of ye church of Eccles, if it be not bylt in my lif, and a rodeseller made, ye some of ten m'ks."

This was on the eve of the great changes in the ritual, applicable to the internal arrangements of our churches. Up to the middle of the sixteenth century, of the rood lofts in our churches, some were most elaborate and costly specimens of composition and wood carving. Several rood-lofts still exist in various churches, especially in Somersetshire and Devonshire. In Long Sutton church, Somersetshire, is a splendid wooden rood-loft, elaborately carved, painted, and gilt, which extends across the whole breadth of the church, and is approached by means of a staircase turret on the south side of the church. These turrets are not unfrequently found attached to large churches on the north or south side near the east end of the aisles or nave. In small churches the stone steps leading up to the rood-loft are still existing near the chancel arch, as in the churches at Brinklow and Newbold-on-Avon, Warwickshire, whilst the door or doorway appears above at some height. Banwell church, Somersetshire, has also a rich rood-loft. At Totnes, Dèvonshire, the upper portion of the screen-work is filled with arched panels of open work, with window tracery and mullions of the fifteenth century, and the rood gallery still remains. There are also in the churches of this county, many stone screens of elaborate workmanship. In the churches of Great Handborough, Enstone, Great Rollwright, and Hook Norton, Oxfordshire, are considerable remains of the ancient rood-loft. Many rich roodlofts were removed from conventual churches on their suppression and destruction to neighbouring parish churches, and there set up. A curious example of this may be seen in Llanwryst church, North Wales, taken, it is said, from a neighbouring conventual church, Macnen, and here set up. And here a peculiarity, which may, at the time of the removal, have been designed, exhibits itself; for the crest or transverse beam, which supported the image of the crucifix and the attendant images of St. Mary and St. John, as appears by the morticed holes, has been placed eastward of the loft, instead of westward.

Besides the rich rood-lofts in large churches, even small churches were not without plain and sometimes even with rich examples, but of very limited dimensions, as at Wormleighton church, Warwickshire, to which loft there is no apparent access; the little church of Coates, near Stow, in Lincolnshire; the little church of Patricio, near Crickhowel, South Wales; and Llaneilian church, Anglesey. In this latter church the rood-loft is tolerably perfect; it has a coved projection on either side of the screen which supports it, and the cornices are carved. It is seven feet in width, and the entrance to it is up a newel-staircase, in the south wall of the nave at the east end. It is but rarely we find in a small church like this the rood-loft in so perfect a state.

Hardly a rood-loft is, however, remaining of earlier date than the fifteenth century; prior to that period, and in many instances even during it, the crucifix or rood and its attendant images appear to have been affixed to a transverse beam extending horizontally across the chancel arch; this was sometimes richly carved; and a beam of this description still exists parting the chancel and the nave of Little Malvern church, Worcestershire. Of the demolition of the rood-lofts generally I purpose to treat in a future chapter.

Of the rood-loft images, out of the general destruction by authority in the reigns of Edward the Sixth and Elizabeth I know of one set only which has escaped. This is in the little church of Bettys Gwerful Goch, near Corwen, North Wales, where the image of the crucifix, of St. Mary, and St. John, rudely carved on a wooden panel in low relief, and formerly affixed to, or in front of the rood-loft, are still preserved and placed as a reredos over the holy table. The panel, four feet and three-and-a-half inches wide, by two feet and three inches in height, is divided into five compartments, each from seven-and-a-half to eight inches wide. The central compartment contains a rude representation, in low relief, of the crucifix, the figure of which is very indistinct; on the sides of the head of the cross are the words "*Ecce Homo;*" on the compartment on the one side next to the crucifix, rudely carved in low relief, is the figure of the Blessed Virgin, in a veiled head-dress, a nimbus over the head, and the hands folden on the breast; by her side, in the outward compartment, are represented the pincers, thorns, and nails. In the compartment on the other side of the crucifix, St. John is represented holding his right hand to his head, and in the compartment beyond this are carved the hammer, the reed with hyssop, like a club and spear. The whole is a specimen of very rude carved work of the fifteenth, or early part of the sixteenth, century. In Great Rollwright church, Oxfordshire, a few years ago part of the rood itself was existing.

On the floor of the tower of Collumpton church, Devon, there is, or recently was, carved in wood, a representation of rock-work with sculls, a calvary, being the base of the rood, with the socket or mortice hole in which the crucifix was fixed.

Dineley, in *An Account of the Progress of His Grace Henry the First Duke of Beaufort, through Wales*, 1684, mentions having seen in Llanrwyst church the wooden image of the crucifix belonging to the rood-loft there, which had been removed, and though kept concealed in the church, was not generally shewn. The following are his words:—" Over the timber arch of the chancell, near the rood-loft, lieth hid the ancient figure of the crucifixion, as bigg as the life. This, I suppose, is shewn to none but the curious, and rarely to them."

Apart from the rood-loft images, we sometimes met with bas-relief sculptures of our Saviour extended on the cross, with a figure on each side representing the Blessed Virgin and St. John, but in a mutilated condition. On the outside of the west wall of the south transept of Romsey church, Hants, and close to the entrance from the eastern cloisters into the church, is a large stone rood or crucifix, sculptured in relief, with a hand above emerging from a cloud:* this is apparently of the twelfth century. Small sculptured representations of the rood, with the figures of St. Mary and St. John, still exist on one of the buttresses near the west door of Sherborne church, Dorsetshire; over a south doorway of Burford church, Oxfordshire; in the wall of the tower of the church of St. Lawrence, Evesham; and on the Norman tympanum of the south door of Bolsover church, Derbyshire.

Passing through the doors of the screen, or beneath the rood-loft which separated the choir or chancel from the nave or body of the church, we perceive in choirs of conventual churches, as also in our cathedrals, and in some parish churches connected in some measure with conventual or monastic foundations, on either side of the entrance facing the east, and also along the north and south sides to a certain distance, a range of wooden stalls divided into single seats, peculiarly constructed,

* "Superest exponere, quod manus illa e nubibus erumpens indicet: Quæ procul dubio omnipotentis Dei dextram designat."—*Ciampini Vetera Monimenta, vol. ij. pp.* 22, 81.

the *formulæ* or forms of which were moveable, and carved on the *subsellia* or under-sides with grotesque, satirical, and often irreverent devices, or illustrations taken from medieval romances; these were appropriated to the monks of the monastery of which the church formed part, or canons of the cathedral in which these stalls are found. The form of each stall, when turned up so as to exhibit the carved work on the under part, furnished a small kind of seat or ledge constructed for the purpose of inclining against rather than sitting on, and this was called the *misericorde* or *miserere*. The *formulæ* or forms when down, and the *misericordes* when the forms were turned up, were used as the time required for penitential inclinations.*

In front of these stalls was a desk ornamented on the exterior with panelled tracery; and over the stalls, especially of those of cathedral churches, canopies of tabernacle work richly carved were often disposed. In Winchester Cathedral we have perhaps the most chaste and beautiful examples of the canon's stalls, with canopies over, that are to be met with, and these are of the fourteenth century, although a greater profusion of minute carved ornament may be found in the canopies which overhang the stalls of later date in other cathedrals. In old conventual churches which, or portions

* In a MS. containing the rules of the Carthusian order, in the Cottonian collection, is the following passage:—"Item tunc stent in sedibus suis versa facie ad altare, donec ad *misericordias* vel super *formulas* prout tempus postulat inclinent, a laudibus enim vigiliæ natalis Domini usque in crastinum octabarum apparitionis et a Pasca in crastinum octabarum Pentecostes et infra octabas corporis. Christi, assumptionis et natalis beate Marie et in festis xij. lectionum ad *misericordias* inclinamus, omni vero alio tempore procumbimus super *formulas*."—*Monasticon, vol. 6, p. v.*

of which, have been suffered to remain, though now no longer used as such, but simply as mere parochial churches, the stalls have been often removed from their original position to other parts of the church. In such conventual churches as were entirely destroyed or laid in ruins, these stalls were frequently removed to neighbouring parish churches, as those in Beaumaris church, Anglesea, from the neighbouring friary church of Llanvaes.

On the carved subsellia of the choir stalls thus noticed, and which are very numerous, I now proceed briefly to treat. The earliest of these may be traced to the thirteenth century, and we find them of all subsequent periods up to the religious changes in the middle of the sixteenth century or nearly so. Some partake of a semi-religious though satirical character, representing wolves and foxes clad in the garbs of St. Francis and St. Dominic, which religious orders somewhat interfered with the secular clergy, as may be seen in the subsellia of the stalls of St. Mary's church, Beverley, where both these orders are represented clad in the peculiar weeds or habits worn by them, for the Franciscan habit differed from that of the Dominican. Rarely indeed in these carvings do we find purely religious subjects treated of in a reverential manner.

The Bestiaries, or books of beasts, a singular compound of fabulous and natural history, as understood in the Middle Ages, furnish subjects for the carvings of many subsellia. The unicorn and virgin, dragons, chimeras, griffins, pelicans, elephants, lions, and monkeys are represented in different attitudes.

The medieval Romances furnished other examples, as that of reynard the fox, rats hanging a cat, a knight in combat with a giant, knights combating wyverns and dragons, and stories of a like nature.

Scenes from domestic life, mostly satirical, are frequent. At Gloucester is a scene representing boys playing at ball; the costume is picturesque—each appears in the *tunica botonata cum caputio*, a dress which denotes the fourteenth century; and indeed the mouldings of the stalls, and the armour or costume in which human figures, male or female, appear, enable us to approximate a probable date to these carvings. Sometimes the schoolmaster is represented in the exercise of discipline on some unruly scholar. Besides the principal or central group, stalks from the cornices of the subsellia bending down support lateral subjects of varied design. On one of these, at Ludlow, is represented a holy-water vat, and aspergillum or sprinkler, a high panelled tomb, a spade and a shovel. Another lateral design at Ludlow represents the mouth of hell.

The subsellia of some stalls in St. Michael's church, Coventry, afford representations, somewhat mutilated, of the dance of death, a burial in which the corpse, enveloped in a shroud, is being lowered into a tomb, and figures in shrouds arising from tombs, denoting the resurrection. These compositions appear to me to be of the early part of the sixteenth century.

The subsellia of the stalls in Exeter Cathedral are perhaps amongst the earliest we possess, and of the thirteenth century. The carvings on many are mainly or altogether of stiff-leaved foliage, simple in design, and formal in execution. Besides these are represented a knight in a hawberk and surcote, in a boat towed by a swan; a merman and mermaid; a knight in a hawberk and sleeveless surcoat, in a cylindrical helmet and heater-shaped shield, attacking a lioness or leopard; a knight, in similar armour, thrusting a sword into a bird; an elephant. All these are of the thirteenth century. Some of the carved subsellia in Worcester Cathedral are amongst the best specimens of art of the fourteenth century, but the most numerous class of wood carvings on subsellia are of the fifteenth century.*

These subsellia are very numerous. In the large parish church of Boston, Lincolnshire, there are no fewer than sixty-four stalls, very few of which have lost their subsellia. The subjects of the wood carving of the latter are very diversified, and do not constitute anything like a series of scriptural, legendary, or other events, but exhibit a mixture of grotesque and miscellaneous compositions.†

Perhaps some of the latest of these subsellia appear attached to the stalls in Henry the Seventh's chapel, Westminster Abbey. These are more or less in the style of the Renaissance, and of the early part of the sixteenth century.

* A detailed Essay on the Sculptures of Misericordes in our Churches was projected upwards of thirty years ago by the late Mr. Thomas Wright and Mr. Fairholt, but I cannot find that the work was ever published.

† These subsellia are fully described by the Right Reverend Edward Trollope, Bishop Suffragan of Nottingham, in Vol. X. of *Associated Architectural Societies' Reports and Papers*, 1870, with sixteen wood engravings executed at the cost of F. L. Hopkins, Esq., of Boston, in illustration.

Church FURNITURE may be described as divided into two classes, moveable and immoveable; the former, including vestments for the clergy, are mentioned under the generic term of ornaments—*ornamenta.* Certain articles were enjoined, by divers Episcopal Constitutions, to be provided at the expense of the parishioners. These were deemed necessary for the proper and efficient performance of Divine service. Other articles, such as organs, not absolutely necessary, were contributed at the cost of private individuals. Some of those pertaining to the altar service were provided at the expense of the rectors or vicars of churches.

In the Excerpts of Ecgbert, Archbishop of York, A.D. 750, each bishop was required to take care that the churches in his diocese were well constructed and kept in a proper state of restoration, and were duly ornamented and provided for both as to the material building, as in lights and otherwise. No altars were allowed to be consecrated except they were of stone and anointed with chrism.

In the Constitutions of Walter de Cantilupe, the celebrated Bishop of Worcester, A.D. 1240, one is '*De ornamentis suarum ecclesiarum*:' of the ornaments of churches. By this it was enjoined that the material building should be kept clean, properly covered, and well attended to; that they should present to sight fitting ornaments; that pertaining to the altar in each church there should be three albs, with amices, and stoles, and maniples, two surplices, and two rochets, two chesibles, two pair of corporals, four of them consecrated and of linen. Two altar palls, two chalices of silver in large churches, and a third of base metal, *stanneus*, unconsecrated, to be carried to the sick. Two pixes, one of silver, or ivory, or of limoge work, *de opere lemonitico*, in which the hosts were to be kept, the other decent and comely, in which the offerings were to be deposited. Two phials, the one for the wine, the other for the water (for mixture in the chalice), a pair of candlesticks, a thurible, a chrismatory fitting and becoming, two crosses, one of them a processional cross, the other for use at the burial of the dead. A banner, a lenten veil, *Sacrariun immobile*,* a lantern and two small handbells, a fitting bier on which the dead were to be carried to burial, and for the use of which nothing was to be charged, and a vessel for holy water.

The service books required for each church were the missal, the breviary, the antiphonary, the gradual, the topary, the manual, the psalter, and the ordinal.

The laity were not to stand in the chancel during Divine service, with the exception of patrons, and personages of rank. After the feast of the Holy Trinity, what remained of the Pascal wax candle was to be converted into smaller candles for the use of the altars and the poor.

By the Constitutions of Walter de Kirkham, Bishop of Durham, A.D. 1255, the Eucharists, which is the Sacrament of the Lord's Body, *quæ est sacramentum dominici corporis*, were to be deposited in a clean and well-known place, and kept under lock; and the priest

---

* This is defined by Du Cange as, "*Pars altaris, ubi reponitur Pixis, in qua sacra Eucharistia asservatur.*"

was to instruct the people, that when, in the celebration of mass, the levation of the host took place, they were to make due obeisance.

In the Statutes of John Peckham, Archbishop of Canterbury, A.D. 1280, is one "*De ornamentis ecclesiæ ad parochianos pertinentibus.*" Of the ornaments of the church to be provided by the parishioners. By this the parishioners were to understand that the chalice, missal, and principal vestment (the latter word being used in a generic and not specific sense) of the church, *viz.*, the chesible, alb, amice, stole, maniple, and girdle, with two towels, the great processional cross, and a lesser one for the dead, a lantern, with a small bell, a thurible, a lenten veil, a banner, hand bells, *campanæ manuales*, for the dead, a bier, a vessel for holy water, a pax, a candlestick for the pascal taper, bells in the belfry and ropes for the same, a consecrated font with a lock, the reparation of the nave of the church within and without, as also the altars, images, windows and glazing, with the enclosure of the cemetery, fell on the parishioners; but all other things, such as the reparation of the chancel and its ornaments, both internal as well as external, according to divers approved constitutions, fell on the rectors and vicars.

In the Synod of Exeter, held A.D. 1287, it was enjoined that the priest about to celebrate mass should pour into the chalice a greater quantity of wine than of water, of which latter there was to be but little; and that masses should not be celebrated except upon consecrated altars, at which celebration there should be at least two lights, as well in reverence of the sacrament as in case of the accidental extinction of either. Of these lights one was to be of wax.'

This Synod also treated, "*De ecclesiarum ornamenta,*" of the ornaments of churches, enjoining that there should be in every church at least one chalice of silver, pure, a cup of silver or pewter, *ciphus argenteus vel stanneus*, for the sick, so that after the reception by them of the Eucharist, the priest might minister, *præbeat*, to himself the washing of their fingers in the same. Two clean and perfect corporals *cum repositoriis*, four towels for the high altar, of which two at least should have received benediction, and one of them with its parure; the same for each altar when it happened that mass should there be celebrated. Two surplices, one rochet, a lenten veil, a nuptial veil, a mortuary pall, a frontal for each altar, a good missal, a gradual, a topary, a good manual, a legendary, an antiphonal, psalters, an ordinal, a collection of hymns. A chest for the books and vestments. A silver or ivory pix with a lock, for the Eucharist. A pewter chrismatory with a lock, a pyx for the oblations, a pax of wood, *asser ad pacem*, three phials, an immoveable stone altar, *sacramentarium lapideum et immobile*, a thurible, a vessel for incense, a vessel for holy water. *Hercia ad tenebras.*' A candle-

---

' The reason why is explained :—Quia vero per haec verba "Hoc est enim corpus meum," et non per alia panis transubstantiatur in corpus (Christi) prius hostiam non levet sacerdos.—Hostia autem ita levatur in altum ut a fidelibus circumstantibus valeat intueri ;......Parochiani solicite exhortentur nt in elevatione corporis Christi non irreverenter se inclinent, sed genua flectant, et creatorem suum adorent omni devotione et reverentia, ad quod per campanellæ pulsationem primitus excitentur et in elevatione ter tangatur campana major.

' Candelabrum Ecclesiasticum in modum occæ, seu trigoni, confectum

stick for the pascal light. Two crosses, the one fixed, the other portable. An image of the Blessed Virgin, and of the local saint. The pascal wax, two processional wax lights. A sculptured table over the altar, *caelatura super altare*. A small bell to be carried (in procession with the host) to the sick, *campanella deferenda ad infirmos*, and at the levation of the body of Christ. A lantern, *boeta?* Small bells for the dead, a bier for the dead, a stone font kept locked, *Baptisterium lapideum bene seratum*. Windows, properly glazed, both in the chancel and nave. *Duo philatoria ad cornu altaris, et unum ad patenam?*

The Constitutions of Robert de Winchelsey, Archbishop of Canterbury, A.D. 1305, treat of the articles pertaining to churches, and point out what the parishioners were to furnish, and what the rectors. The parishioners were to provide books, *viz.*, the legendary, the antiphonar, the gradual, the psalter, the tropary, the ordinal, the missal, the chalice; the principal suit of vestments, *vestimentum principale*, with a chesible, dalmatic, and tunic, a choral cope, with all appendages; a frontal for the high altar, with three towels, three surplices, one rochet, a processional cross, a cross for the dead, a thurible, a lantern, a small handbell, *tintinnabulum*, to carry before the host, *coram corpore Christi*, in the visitation of the sick. A fitting pix for the host, a lenten veil, a banner for rogations, bells with ropes, a bier for the dead, a vessel for holy water, a pax, *osculatorium*, a candlestick for the pascal wax taper, a font with a lock.

nostris Herce.—candelabrum in herciæ modum confectum luminibus variis instructum, quod ad cenatophii caput erigi solet.—*Du Cange.*

Images in the church; the principal image in the chancel; the inclosure of the cemetery; the reparation of the body of the church, both inwardly and outwardly, as well as in images as in glass windows; the reparation of books and vestments, whenever such was required. But as to other things, as in the reparation of chancels, according to divers customs, the costs of reparation fell on the rectors and vicars of the different places.

In the "Concilium Provinciale Cashelense," provincial Council of Cashell, in Ireland, held A.D. 1453, it was enjoined that in every church there should be at least three images, namely, of St. Mary the Virgin, of the crucifix, and of the patron of the place, in honour of whom the church was dedicated.\*

But besides the images thus specially enjoined and required to be placed in every church at the expense of the parishioners, many other images of saints, or such as were so esteemed, were made at the costs of and presented by individual benefactors, or left by will to churches; and the brackets on which they were placed are still retained, mostly projecting from one side or both of an east window, and these are very numerous. Sometimes we find a bracket forming the base of a rich canopied recess, within which the image was set, as at the east end of the north aisle of Monks Kirby church, Warwickshire, and elsewhere.

William, Earl of Salisbury, by will dated 20 April, 1397, willed that xx*s*. be disposed of to make an image

---

\* Quod in singulis ecclesiis ad minus habeantur tres imagines, sanctæ beatæ Mariæ virginis, sanctæ crucis, et patroni loci in cujus honorem ecclesia dedicatur.

of St. Ann of alabaster, to be placed on the altar of the Blessed Virgin at Henyngs.*

William Bruges, Garter King of Arms, London, by will dated 26 February, 1449, bequeath to the chapel of our Lady, St. George's church, Staunford, two images of our Lady and St. George, being in painted stone; and in his chapel at Kentishton, and to the chapel of St. George, of Staunford, the ymage of the Trinity of stoon standing in his chapel at Kentishton.

Richard Tylle, of Sellyinge, by will dated 17 December, 1485, bequeathed to the making of a new image of our Lady in the church of Sellying, lxvis. viii*d*.

Eleanor Dutchess of Bucks, by will dated 24 June, 1528, "I will that my heart be buried in the church of the Grey Friars, London, before the image of St. Francis." Such are amongst a multitude of similar bequests.

The introduction of IMAGES into Christian churches, with their subsequent adoration, a practice perhaps earlier than the sixth century, but the actual commencement of which it is now difficult to ascertain,ʸ was noticed

---

* This and the following extracts from Wills are taken from *Nicholas' Testamenta Vetusta*.

ʸ Irenæus, who wrote in the latter part of the second century, in his exposition of the Carpocratian heresy, has the following remarkable passage, the Greek text of which is lost, but is repeated nearly verbatim by Epiphanius about the latter part of the fourth century. " Unde Marcellina quæ Romam sub Aniceto venit cum esset hujus doctrinæ multos exterminavit. Gnosticos se autem vocarit, et imagines quasdam quidem depinctas, quasdam autem et de reliqua materia fabricatas habent, dicentes formam Christi a Pilato illo in tempore quo fuit Jesus cum hominibus. Et has coronant et proponunt eas cum imaginibus mundi Philosophorum; videlicet cum imagine Pythagoræ, et Platonis et Aristotelis, et reliquorum, et religiosam observationem circa eos similiter ut Gentes faciunt."—*Irenæi op. l. 1. contra hæreses*.

St. Augustine also, enumerating the heretical doctrines of Carpocrates,

by Pope Gregory the Great, at the close of that century, A.D. 598. He indeed censured their worship, though he maintained they might be of use, especially to the unlearned, as incentives to devotion, and as vehicles of instruction. After the death of Gregory, in the early part of the seventh century, image worship became prevalent in the west, and so continued to the close of that century. About the year 726 the Emperor Leo III. published an edict prohibiting the religious adoration of images. In the middle of the eighth century, A.D. 754, a Council of Constantinople condemned the religious use of images. A.D. 775 image worship was opposed by the Emperor Leo IV. About A.D. 784 the Church of Rome declared in favour of image worship. By the seventh General Council of Nice, held A.D. 787, religious adoration of images was established.

Charlemagne maintained, A.D. 790, that no kind of veneration ought to be paid to images, though they might be retained in churches as incentives to devotion; and, A.D. 794, a Council held at Frankfort decided in favour of the views of Charlemagne respecting image worship. A.D. 815, a Council held at Constantinople forbad the worship of images. The Emperor Theophilus, A.D. 830, published an edict prohibiting all use of images in churches.

thus expresses himself: "Sectæ ipsius fuisse traditur quædam Marcellina quæ colebat imagines Jesu et Pauli et Homeri et Pythagoræ, adorando incensumque proferendo."

In the beginning of the third century the Emperor Alexander Severus is said to have had the images of Abraham and Jesus Christ placed together with those of Orpheus, Apollonius and other deities in the place where he performed his daily devotions.

Such passages, if correct, would tend to prove the fabrication of images of our Saviour, so early as the second century.

A Council held at Constantinople, A.D. 842, confirmed the decree of the seventh General Council with regard to images, and they were restored to churches.

Image worship is supposed to have been established in England towards the close of the ninth century.

Godiva, wife of Leofric, Earl of Mercia, who died about the middle of the eleventh century, circa A.D. 1050, shortly before her death gave a chain of precious stones to be placed round the neck of the image of the Blessed Virgin, in the church of the monastery at Coventry, founded by Earl Leofric and herself, so that those who came of devotion thither should say as many prayers as there were gems therein.[*]

The destruction and removal of images from our churches in the middle of the sixteenth century will be treated of in a future chapter; the remains of such, more or less mutilated, have been from time to time brought to light.

The earliest sculptured image I have met with in this country is one discovered in the east wall of the choir, York Cathedral, after the fire of 1829. It is in relief, and represents the Blessed Virgin, sitting, with the Divine Infant in her arms. The heads of both Mother and Child have been struck off, and the work otherwise mutilated. The Virgin appears clad in a gown and mantle, and the folds of the drapery evince it to have been executed in the Norman period, *viz.*, in the twelfth century. The size I am unable to determine.[a]

In a niche over the entrance into the porch, which is

[*] William of Malmesbury.
[a] Engraved in *Poole and Hugall's Historical and Descriptive Guide to York Cathedral.*

of the decorated style, of Horley church, Oxon, is an image of alabaster, in bas-relief, apparently intended to represent St. Margaret with a dragon at her feet. This may have been intended for the "*principalis imago.*"

In the ancient domestic buildings adjoining the site of the Abbey church of Muchelney, Somersetshire, now destroyed, is a very perfect but small alabaster image of St. Michael, which I imagine to have been the "*principalis imago*" of a church. This was dug up in the churchyard of Othery, in the neighbourhood.

In the cloisters of Lincoln Cathedral is an image of St. Giles, brought from the ruins of the Hospital of St. Giles, which stood north-east of the cathedral, outside the close. It represents the Saint, vested as an ecclesiastic of priestly rank, in the alb with its girdle, and the stole crossed in front of the breast, with the extremities hanging down on each side; about the neck appears the amice with a rich apparel or parure, and over all a cope is worn. The arms are somewhat mutilated. At the feet reposes a mutilated animal, apparently a hind, from the hip to the shoulder of which, and right across the flank is an arrow. Such is the symbol originating from the legend of St. Giles, who is said to have lived in the latter part of the seventh century. It was probably the "*Imago principalis*" placed in the chancel of the chapel of the hospital, and as such subject to veneration, Dulia.

When the images were removed from our churches by authority in the middle of the sixteenth century, the compulsory office fell upon some who had little inclination for the task, and who buried the sculptured image in the churchyard, face downwards, so as to effect as little injury as possible. More than one image has been discovered thus concealed.

In the restoration of Breadsall church, Derbyshire, in 1877, on removing the pavement under the west gallery, at the south-west corner, the workmen came across a large stone buried beneath the surface of the soil; on being taken up, a piece of finely developed sculpture was disclosed, representing our Lady of Pity with the dead Christ upon her knees. She is represented in a gown and mantle, with a coverchief on the head, and the neck bare; the costume being referable to the early part of the fifteenth century. This image shows indications of colour and gilding. The entire height is two feet five inches, and the entire breadth one foot five inches.

In Battlefield church, Salop, is a wooden image of "*Our Lady of Pity.*" This is three feet nine inches in height, carved out of a block of oak, hollowed behind; the Blessed Virgin is represented in a sitting attitude, supporting on her knees and in her arms the dead body, nude, with the exception of a cloth about the loins, of our Saviour. She is attired in a gown with ample skirts, and mantle, and her head is covered with a veil or coverchief falling down behind. The execution of this image is good, but such as clearly indicates it to have been executed in the fifteenth century. In general conventional design it resembles the stone image found in Breadsall church.[b] About the middle

[b] The wooden image in Battlefield church is engraved in the 14th Vol. of the *Archæologia*. The stone image discovered in Breadsall church is engraved in the 34th Vol. of the *Journal of the British Archæological Association*.

of the eighteenth century there appears to have been another wooden image preserved in this church, which was destroyed when the church underwent reparation.

To these images of our Lady of Pity excessive reverence appears to have been formerly paid, for in "a goodly Prymer," being the first of three Primers put forth in the reign of Henry VIII., and published A.D. 1535, in "An Admonition to the Reader," the images of our Lady of Pity are thus alluded to: "What vanity is promised in the superscription or title before *Obsecro te, Domina Sancta Maria?* where it is written that whosoever saith that prayer daily before the image, called the image of our Lady of Pity, shall see the visage of our most Blessed Lady, and be warned both of the day, and also of the hour of his death, before he depart out of this world." . . . . . . .

"But what blindness is that to appoint the prayer to be said before the image of our Lady of Pity? I pray you what and if a man did use to say it before the image (as they call it) of our Lady of Grace? Shall he then lose the inestimable privileges before promised? Yea I pray you why might not a man smell a little idolatry here, in that there appeareth in this title a certain respect, a reverence more to one image than to another? Men will say they honour no images, neither of stocks nor yet of stones, and that neither man, woman, nor child is so mad so to do: and yet must this prayer be appointed to be said before the image of our Lady of Pity, in a manifest and open token and sign of a peculiar honour and reverence to be done to the same image or picture."

Wooden images are now of great rarity. In a room adjoining St. Mary's Hall, Coventry, is a wooden image of St. George, representing him on horseback combating the dragon; the armour in which he is clad being of that kind in use in the reign of Henry VI. From the flat surface on one side, this image appears to have been affixed against the wall, and'was placed over the altar in the chapel of St. George, adjoining one of the city gates, Coventry. This carving is two feet six inches in height.

There is another well-known wooden image in Coventry which must have been removed from one of the churches, and was probably "a George." From the details, viz., the broad-toed sollerets, it appears not older than the reign of Henry VII.[e]

I have in my possession a small wooden image of a Bishop, enshrined, as it were, within tabernacle-work. This is of the early half of the fourteenth century, and was formerly in the church of Dunchurch, Warwickshire, rebuilt by the monks of Pipewell, in the fourteenth century. It was thrust out of that church at the commencement of the present century, when the finely-carved oak sittings were displaced and swept away for the incoming of wretched deal boxes. The effigy of the Bishop is vested in alb, stole, dalmatic and chesible, with the mitre on the head, the amice about the neck, and the maniple over the left arm. It is in height three feet three inches, but whom it represents I know not.

[e] For the last two centuries this appears to have done duty as "Peeping Tom" of fabulous notoriety, as connected with the fabulous legend appropriated, as elsewhere, to this city.

In Abergavenny church, South Wales, is a large wooden image, apparently that of Jesse, in a reclining position. This appears as if part of a design.

In Llandderfel church, North Wales, is still existing a rudely carved wooden horse or animal, an appendage to the famous image of the patron saint of the church St. Dervel Gadarn, which was taken up to London in 1538, and consumed in the fire at Smithfield, with which Friar Forest was burnt for denying the king's supremacy.

In Sandford church, Oxfordshire, is an image representing the assumption of the Blessed Virgin. This was discovered in 1723 with the sculpture downwards, having been so placed for preservation, and having the appearance of a plain flat stone. The Blessed Virgin is represented clad in a gown and mantle, with numerous folds, with a crown upon her head, and surrounded with a halo of rays; the hands are gone. Six angels appear: three on each side. Beneath the image is a small tabernacle, supported by two angels, and apparently intended to receive the pix, with the host inclosed. From the high-bowed crown this image cannot, I think, be of earlier date than the reign of Henry VII.[d]

Dispersed about in Hereford Cathedral are certain images not occupying their original positions. One of these is of St. John the Baptist, clad in a tunic and mantle, with the *Agnus Dei*, and in the left hand a cross. Another is of an Archbishop or Pope, holding a double crozier, the staff of which is veiled round. The prelate

[d] It is engraved in *Skelton's Oxfordshire*.

thus represented is vested in the alb, tunic, dalmatic, and chesible, with the mitre on the head. This is apparently a production late of the fifteenth, or early in the sixteenth, century. A third is that of a headless image, holding in the left hand a book. A fourth is the image of a Bishop, wearing on his head the *mitra pretiosa*, and vested in the alb, tunic, dalmatic, and chesible. His right hand is upheld in act of benediction, whilst his left hand grasps the pastoral staff.

Such images must not be confounded with mere decorative imagery or effigies on the exterior of our churches, or as accessories to tombs.

A bracket projecting from one of the piers of Durham

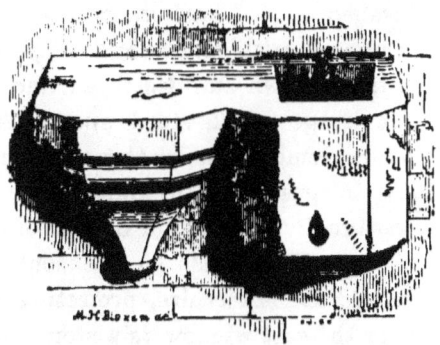

Pixis, or Ancient Stone Offertory Box and Image Bracket, Bridlington Priory Church, Yorkshire.

Cathedral still supports a mutilated image. Projecting from the south-easternmost pier of the nave, Bridlington Priory church, Yorkshire, is a stone bracket for an image; and contiguous to this, and forming part of the design, is a pixis or ancient stone offertory box. The cavity of this box is slanting, and indications are clearly apparent of its having had a cover of wood and a lock. These, however, are gone.

This church was one of those conventual churches, of which there were several, divided into two parts, the one part for the priory, and the other part for the parish. As the shrine of St. John of Bridlington was in the eastern part of the church appropriated to the priory, and between the high altar and east window, all of which has been demolished, this could not have been the offertory box at the shrine, but was probably an offertory box placed by the side of an image of St. John of Bridlington, in the part near that high altar, to which the parishioners had free access. By the Royal Injunctions, issued in 1547, such images as were known to have been abused with pilgrimage *or offerings* were ordered to be taken down and destroyed, and it is probable that the image which stood upon this bracket was, if not before, removed under these injunctions.

Before these images lamps or lights were formerly set or kept continually burning, in honour of the Saint whose image was displayed. In the south aisle of Evington church, Leicestershire, projecting from the north wall, near the east window, is a stone bracket, on which an image formerly stood; and in front of this is a smaller bracket, projecting from the larger, in which is sunk an orifice or socket for a taper or light to be set in. This is a singular example now remaining.[c]

[c] By will dated A.D. 1399, Sir Philip Darcy, Knight, left as follows:— "Item lego XX*s*. ad emendum quandam ymaginem Sanctæ Annæ matris beatae Mariae de alabaustre, ad ponendum ad altare beatae Mariæ de Henynges. Item durante termino quinque, quolibet anno, quatuor libras

## PREVIOUS TO THE REFORMATION.

Projecting from the north wall of the little chapel of Upton Castle, Pembrokeshire, is a man's fist, sculptured in stone, with a perforation for a light or taper to be placed in. Westward of this is a sepulchral effigy.

In the south aisle of Barneck church, Northamptonshire, projecting from one of the piers is a bracket for an image.

Besides the images thus noticed there were in many of our churches sculptured representations, in relief,

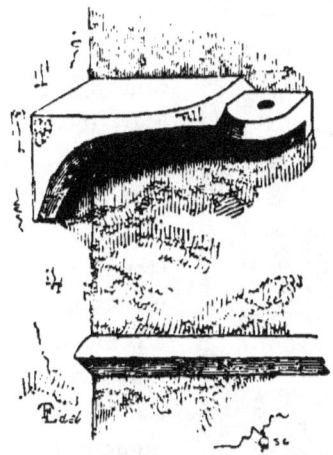

Bracket, Evington Church, Leicestershire.

mostly of Scriptural subjects, known as "tables," and placed over or at the back of altars as a reredos. We carry these back to the twelfth century. In Chichester

ceræ ad comburendum, coram prefata ymagine in honorem Dei, beatae Mariae, et beatae Annae." In the accounts of St. Michael's church, Cornhill, London, between A.D. 1456 and A.D. 1475, are the following entries:—
"Pay$^d$ to West, founder, for amendyng of a candlestyk afore Saynt Barbara, VIII$^d$." It'm for makyng clene of the candelstyk afore Seint John, iiij$^d$.

Cathedral are two sculptured slabs of this description, discovered in 1829 behind the stalls of the choir, where they had been concealed. The subjects were the raising of Lazarus, and the meeting of our Lord with Martha and Mary. The treatment is that of essentially the twelfth century, the folds of the drapery of each figure being numerous and conventional. They are said to have been removed from Selsey.

Norman Sculpture, Bobbing Church, Kent.

In 1863, whilst repairs were taking place in the parish church of Bobbing, Kent, a piece of oolitic stone, about two feet four inches in length, and about six inches in width, was found in the south wall of the chancel, forming the quoin of the western jamb of the sedilia, which appears to be of the fourteenth century.*  This stone was sculptured on two sides with figures ten inches in height. One of these figures represents a Bishop, with the low bifed mitre on his head, vested in

---

* This discovery is noticed, and a full description of it given, in the 21st Vol. of the *Journal of the Royal Archæological Institute*.

the alb, with its parure, stole and chesible, the right hand being raised in act of benediction, whilst the left hand grasps the pastoral staff; one extremity of the stole only is visible, beneath the middle of the chesible. The head is surrounded by a halo, indicative of saintly character. Over the head is an inscription L.S.MARCIAL' PĬ.PAT̊NVS. The other portion represents a Bishop in the same vestments, with his right hand upheld in benediction, and his left grasping the pastoral staff; close to this is sculptured a figure with a book in his hand, supposed to be a deacon. This sculpture may be referred to the middle of the twelfth century. When discovered, this stone was placed with the sculptured figures downwards. St. Martial, sent to France from Rome, is said to have flourished in the middle of the third century, circa A.D. 250.

Norman Sculpture, Bobbing Church, Kent.

In Bolsover church, Derbyshire, on the north side of the chancel, and now affixed to the wall, is a "table," sculptured in bold relief, representing the Nativity. In it are introduced the Blessed Virgin and infant Christ, St. Ann and St. Joseph, with the heads of the ox and the ass, which are thus conventionally represented. This table is five feet two inches in width by two feet ten inches in height. It is said to have been found buried in the churchyard, with the sculptured face downwards. It probably formed the reredos to an altar.

In the wall at the east end of Cookhill chapel, Worcestershire, is a sculptured table of the fifteenth century, representing the assumption of the Blessed Virgin. She is represented crowned. On each side of the aurora are four small figures of angels, and beneath are two kneeling figures.

In Long Melford church, Suffolk, is a "table," two feet two inches in width by fourteen inches in height, representing, sculptured in relief, the adoration of the Magi. The Blessed Virgin is represented holding the infant Christ, who is accepting the gifts of the three kings; they appear with crowns on their heads; Joseph is represented with a carpenter's tool; behind the figure of the Virgin appears a female figure, with a veil on the head.

A sculptured "table" in Yarnton church, Oxfordshire, is divided into six compartments, in one of which is represented our Lady of Pity, with the dead Christ in her lap, in the same conventional style as the images at Battlefield church and Breadsall church. In other compartments appear the betrayal of our Saviour, the carrying of His cross, and His crucifixion. This, the gift of the late Alderman Fletcher, of Oxford, is now appropriately placed as a reredos at the back of the communion table.

In Cuberly church, Gloucestershire, a portion of a "table" is preserved, lying loose in the chancel, representing the Rood, St. Mary, and St. John.

In 1458, money was bequeathed "ad novam tabulam de alabastro de historia Sanctæ Margaretæ," in the church of Dunwich, in Suffolk. Four marks were bequeathed to buy a "table" of alabaster, of nine female saints, in St. Peter's church, Norfolk.[g]

In 1510 Robert Clerk wills to be buried in the church, "and a *table* of St. Thomas of Ynde, which I have caused to be made; I will have it stand in Batfield church, Norfolk."[h]

Of the vestments in which images were sometimes arrayed, we have an account in "A list of the goods of Great St. Mary's church, Cambridge, taken the nineteenth year of Henry VII. A.D. 1503," and which contains the following items:—

"It a cote of touvney damaske purfullyd w$^t$ ffelewet, apperteynyng to our Lady."

"It a rede sateyn coote with two payer of beds of blakke geat, apperteynyng to her Sonne."

"It a coote for her Sonne of the same satoyn purfilled w$^t$ blakke velvet w$^t$ spangills of golde."

"It a shoo of silver for the ymage of our Lady."

[g] *Archæologia*, Vol. xii., p. 94.
[h] *Ibid.*

"It a coler of gold ffor to hange abowght oure Ladies nekke off ix lynks in the coler."

Early in the reign of Edward VI., when many of the church goods were sold, the following items appear:—

"1547, Rec$^{d.}$ for certaine old implyments, as paynted clothes, pattyne, candyllstyks, wood ymages, and a barnakyll, solde by the assent of the parish, 22s."

"1550, It sold the rede cote and y$^t$ St. Nycholas dyd were, the coler rede, to James Radclyff, 6s."

"It sold the vestement and cope y$^t$ Seynt Nycholas dyd were."

Amongst an Inventory of church goods, in the parish church of Wolverhampton, taken A.D. 1541, we find mentioned:—

"S$^t$ Catherine's cote of black velvet."

"It' our Ladye's cote of black velvet, and two sleeves, one of red tynsell satten, and the other of black velvett, and Ihs cote of red sattin."

Amongst the Articles devised by the bishops for Master Latimer to subscribe unto (A.D. 1531), are the following:—"That it is laudable and profitable that the venerable images of the crucifix and other saints should be had in the churches as a remembrance, and to the honour and worship of Jesus Christ, and his saints." "That it is laudable and profitable to deck and to clothe those images, and to set up burning lights before them, to the honor of the said saints."[i]

In the choirs of cathedral and conventual churches, and in the chancels of some other of our churches, a

---

[i] *Fox's Martyrology.*

moveable DESK, at which the epistle and gospel were read, was placed. This was often called the eagle desk, from its being frequently sustained on a brazen eagle with expanded wings, elevated on a stand, emblematic of St. John the Evangelist. Eagle desks are generally found either of the fifteenth or seventeenth century; notices of them occur, however, much earlier. In an account of ornaments belonging to Salisbury cathedral, A.D. 1214, we find mentioned *Tuellia una ad Lectricum Aquilae*. In 1249 John de St. Omer was to make a *Lectron* for the chapter house at Westminster, after the design of that in the chapter house at St. Albans, or more beautiful if it could be.[k] And in the Louterell Psalter, written circa A.D. 1300, an eagle desk, supported on a cylindrical shaft, banded midway down by an annulated moulding in the style of the thirteenth century, is represented. Besides the brass eagle desks which still remain in use in several of our cathedrals, and in the chapels of some of the colleges at Oxford and Cambridge, fine specimens are preserved in Croydon church, Surrey; in the church of the Holy Trinity, at Coventry; in St. Mary Ottery church, Devon; in St. Peter's church, St. Albans; and in Oundle church, Northamptonshire. In Cropredy church, Oxfordshire, a large brass eagle, of ancient workmanship, is used as a support to the modern reading-desk. In Leighton Buzzard church, Bedfordshire, is an ancient wooden eagle desk of the fifteenth century.

There was formerly an eagle of brass in St. Michael's church, Coventry, which was given A.D. 1359, by William

[k] Rot. Claus., 33 Henry III.

Boltoner, and stood at the little quire just at the entrance into the chancel, and in the centre of the aisle. This, in 1645, was condemned by the Puritans, and sold for old metal. "1645, Rec<sup>d.</sup> of Chamberlayne Smith for y<sup>e</sup> Eagle and foote sold by order of vestry for 5<sup>d</sup> the lb. £8. 3*s.* 6*d.*" Consequently it weighed 392 lbs. The brass eagle desk now preserved in the Collegiate church at Southwell, in Nottinghamshire, was about the middle of the last century recovered from a deep part of the lake at Newstead.¹ Being sent to a clockmaker to be cleaned, he found it composed of different pieces; on unscrewing these, the boss on which the eagle stood was found to contain a number of deeds and parchments relating to Newstead Abbey.

According to Du Cange, the eagle designated St. John the Evangelist. The same writer records the bequest of one to the church of Ambrose so early as A.D. 1014.

The author of *A description or Briefe declaration of all the Ancient Monuments, Rites and Customes, belonginge or beinge within the Monasticall Church of Durham before the Suppression,* written A.D. 1593, notices as follows:—
"At the north end of the high altar there was a goodly fine letteron (lecturn) of brasse, where they sunge the epistle and the gospell, with a gilt pellican on the height of it, finely gilded, pulling her bloud out of hir breast to hir young ones, and winges spread abroade, wheron did lye the book that they did singe the epistle and the gosple. It was thought to bee the goodlyest letteron of brass that was in all this countrye. It was

¹ This is engraved in the 2nd Vol. of *Shaw's Dresses and Decorations.*

all to bee taken in sunder with wrests, every joint from other."

Sometimes we meet with ancient brass reading desks which have not the eagle in front, but both the sides are sloped so as to form a double desk; of these examples

Brass Reading Desk, Merton College Chapel, Oxford.

may be found in Yeovil church, Somersetshire, where the sides of the desks are incised with the demi-figure of a man, with an inscription; and in the chapel of Merton college, Oxford. In both of these the shaft or stem is of the usual design, partly angular, partly cylindrical, and beneath or projecting from the base of each, are four lions.

Ancient wooden reading desks, either single or double, are also occasionally found; some of these are richly carved, others are comparatively plain, but all partake, more or less, of the architectonic style of the age in which they were severally constructed, and from which their probable dates may be ascertained. In Detling church, Kent, is a very fine double reading desk of wood, richly carved with decorated tracery and other detail; the upper mouldings of the stem on which this desk is faced are good, and small trefoil-headed panels surround it. The base projects on all sides, with buttressets at each angle. This desk is five feet five inches in height, and presents a chaste, rich, and beautiful appearance. As a specimen indeed of church furniture of the middle of the fourteenth century, circa A.D. 1350, it can hardly be surpassed. In Bury church, Huntingdonshire, is a wooden desk with a single slope, and the vertical face presented in front is covered with arches and other carved ornament; this perhaps may be referable to the latter part of the fourteenth century. A rich double desk, of somewhat later date, with the shaft supported by buttressets of openwork tracery, is preserved in Ramsey church, Huntingdonshire.

In Aldbury church, Hertfordshire, is an ancient double lecturn or reading desk, of wood, of the fifteenth century, much plainer in design than those at Bury and Ramsey; the shaft is angular, with small buttressets at the angles, and with a plain angular-shaped moulded capital and base, which latter is set on a cross-tree. In Hawstead church, Suffolk, is a wooden desk, with little ornament, supported on an angular shaft with an embattled capital and moulded base, with leaves carved in relief; this is apparently of the latter part of the fourteenth century. In Longfield church, Surrey, is an ancient wooden lectern, consisting of a double desk, the sides of which are pierced; this is sustained on a shaft, with slender buttressets annexed. The ancient wooden desks found in some of our churches must not, however, be confounded with a more numerous class constructed and used after the Reformation.

In Middleton or Long Parish church, Hants, is an eagle desk of wood; the shaft supporting it is panelled. In the choir of Winchester cathedral is a wooden eagle on a stand of the same material, both gilt.

In Bangor cathedral are, or were, the remains of an old wooden reading desk, with carved back and Gothic form, of the early part of the sixteenth century.

The east end of the north aisle of Swanscombe church, Kent, formerly contained a portable desk of oak, on a stem fluted and curiously carved with various crosses and Gothic roses. Whether it is still in this church I know not.

Sometimes we meet with reading desks of stone, but they are rare. Near to the pulpit, and close to the entrance of the chancel of South Burlingham church, Norfolk, is a stone reading desk. In the Priors' chapel, Wenlock Priory, Salop, is an ancient stone reading desk covered with sculptured foliage and other detail in relief. This appears to be early in the thirteenth century; and in Crowle church, Worcestershire, is a stone or marble reading desk. This was dug up in the early part of the present century, near the site of the abbey church of Evesham. It is formed out of a block of white marble, richly sculptured in relief on the four sides. In front is sculptured the presumed representation of St. Egwin, Bishop of Worcester, founder and also Abbot of Evesham Abbey. He appears with the right hand upheld in act of benediction, with the left hand grasping the pastoral staff. The rest of the sculpture is of the stiff-leaved foliage of the description prevalent in the early part of the thirteenth century; and this sculpture agrees in date with the recorded fact that Thomas de Marléberg, the thirty-eighth Abbot of Evesham, in or about the year 1218, returning from Rome, after two years was elected Sacrist. He then made a reading desk behind the choir, which the church had not before, and appointed stated readings to be held near the tomb of St. Wilsius."

HIGH ALTAR. Proceeding up the chancel of a parish church, or, if a cathedral or conventual church, the choir, we ascend by three steps the platford at the east end on which the high altar formerly stood. This was so called to distinguish it from other altars, of which there were often several, in the same church. At this high mass was celebrated, whereas the other altars were chiefly used for the performance of low or private

---

" Secundo vero anno reditus a curia (Romana), factus est sacrista, et fecit lectricium retro chorum, quod prius non erat factum in ecclesia Eveshamensi; et legebantur lectiones juxta tumbam S. Wilsini.—*Nash's Worcestershire,* Vol. i. p. 419. This desk is engraved in the 17th Vol. of the *Archæologia.*

masses, chiefly that, in ancient funeral rites, known as *missa pro defunctis.*

The most ancient altars were of wood, afterwards they were constructed of stone; those of the primitive British church are spoken of by St. Chrysostom. By a decree of the Council of Paris, held A.D. 509, no altar was to be built but of stone. Amongst the excerpts of Ecgbert, Archbishop of York, A.D. 750, was one that no altars should be consecrated with chrism but such as were made of stone; and by the Council of Winchester, held under Lanfranc, A.D. 1076, altars were enjoined to be of stone. The customary form of such was a mass of stone supporting an altar-table or slab, generally six inches in thickness, with the lower three inches chamfered or bevelled, and with five crosses incised on the top of the slab. Thus it was held to resemble the tombs of the martyrs, at which the primitive Christians held their meetings; and it became customary to inclose in every altar relics of some saint, and without such relics an altar was esteemed incomplete. Though many of the ancient chantry or minor altars may be found still existing in their original positions, I only know of two ancient high altars of stone in the churches of this country which have not been removed; these are, the high altar in the church of Arundel, in Sussex, which contains no less than four ancient altars of stone. The high altar in Forehampton church, Gloucestershire, is also existing; the altar table or slab is supported by stone shafts, and not by a mass of masonry. Of chantry, and other altars, notice will be subsequently made.

Two principal causes operated in the progressive development in the internal arrangements of our churches during the Middle Ages. These were the great Eucharistic controversies respecting the Holy Sacrament of the Body and Blood of Christ, in the ninth and eleventh centuries; the other, the rise and gradual development of the doctrine of Purgatory. Other minor causes also assisted in these changes. Anciently the furniture of the altar was confined to the altar-cloth and frontel, and to the vessels thereon necessary for the celebration of the Eucharistic service, *viz.*, the chalice and paten covered with the corporal, neither cross or candlesticks being placed thereon. The representation here given, taken from an illuminated manuscript in the Bodleian, of the early part of the thirteenth century, represents a priest celebrating the office for the dead. *Missa pro defunctis.* In this, vested for mass, the priest appears before an altar, on which is placed a chalice, covered with a corporal, and a missal; but neither cross or candlesticks appear. Behind are the bodies of two priests, uncoffined, as was then generally the case, clad in ecclesiastical vestments for burial, and placed on a bier; behind these is the funeral light, and the cross enjoined to be used at the burial of the dead.

The first great controversy about the Eucharist is said to have originated with Paschasius Radbertus, a monk of Corby, in France, in the early part of the ninth century. He is said to have propounded a doctrine which afterwards developed itself into that of a carnal or corporal presence in the mystical elements. In this, considered as a novel opinion, he was opposed by

Missa pro defunctis; from M.S. 13th Century.

Scotus, Bertramus, and other divines of that age. In the middle of the eleventh century, Berengarius wrote against the opinions which had been promulgated by Paschasius, and raised in the church the second great controversy respecting the Eucharist, being opposed, amongst others, by Lanfranc, subsequently Archbishop of Canterbury, who wrote *De Eucharistiæ Sacramento Liber*, to which Berengarius replied in his work, *De Sacra Cœna Liber*. It was after this second controversy that the levation of the host as now practised by the Church of Rome was insensibly introduced."

Early in the thirteenth century, at the fourth Lateran Council, held A.D. 1215, under Pope Innocent the Third, the doctrine of Transubstantiation was declared to be an Article of Faith.°

When the cross was first placed on the altar it is difficult to say*; the two lights on the altar are supposed by some to have been introduced early in the thirteenth

* De Vert, in treating of this practice, observes: "L'elevation des sacrez Symboles (introduite seulement vers le milieu du xii. siecle.) Cette pratique s'est etablie d'une maniere si lente et si insensible comme on pourra aussi le montrer quelque jour." . . . . . . . . —*Explication des Ceremonies, &c., tome i. p.* 250.

° "Una vero est fidelium universalis Ecclesiæ, extra quam nullus omnino salvatur. In qua idem ipse sacerdos et sacrificium Jesus Christus, cujus corpus et sanguis in sacramento altaris sub speciebus panis et vini veraciter continentur, transubstantiatis pane in corpus, et vino in sanguinem potestate divina, ut ad perficiendum mysterium unitatis accipiamus ipsi de suo, quo accepit ipse de nostro. Et hoc utique sacramentum nemo potest conficere nisi sacerdos, qui fuerit rite ordinatus secundum claves Ecclesiæ, quas ipse concessit apostolis, et eorum successoribus Jesus Christus."—*Decreta Generalia Concilii Lateranensis Tempore Innocentii Papæ III. De fide Catholica.*

? In the *Statuta Synodalia Ricardi* (de la Wich secundi) *Cices. Epis.* occurs the following passage,—" Celebret Sacerdos cruce anteposita." This was A.D. 1216. The altar lights are not mentioned, only the cross.

century by Pope Innocent the Third, at or about the time of the fourth Lateran Council.¶

In the sculptured tympanum of the south portal of Amiens Cathedral, a work of the thirteenth century, an ancient altar is represented whereon appears, between two lights, a plain Latin cross. Subsequently, but not, I think, before the fifteenth century, the crucifix appears to have superseded the cross: this again at a later period was accompanied with the images of St. Mary and St. John. In Sir William Dugdale's *Antiquities of Warwickshire* is an engraving representing the combats of John de Astley and Peter de Masse, and Sir Philip Boyle. This, taken from a painting, from the costume evidently executed in the early part of the sixteenth century, gives us in one of the compartments an altar with its cloth and frontal, where, between two lights,

¶ Innocentius III. Pontifex Romanus. De sacro altaris mysterio, cap. xxi. De candelabro et cruce quae super medio collocantur altaris. Ad significandum itaque gaudium duorum populorum de nativitate Christi laetantium, in cornibus altaris duo sunt constituta candalebra quae mediante cruce, faculas ferunt accensas. Lumen autem candalebri fides est populi. Inter duo candelabra in altari crux collocatur media.—*Patrologie Tom ccxvii. accurante J. P. Migne.*

Mais les chandeliers etaient ils ancienement places sur les autels pour y servir d'ornement?

Il est facile de repondre à cette question; it suffit de rappeler ce que nous disons dans l'article Autel.

Celui-ce etait, exclusivement, destiné à porter ce qui etait indispensable pour le Saint Sacrifice. Quand le Célébrant se rendait a l'autel pour y célébrer, les acolytes portaient les chandeliers, qu'ils tenaient pendant la cérémonie ou qu'ils posaient sur les marches par lesquelles on montait à l'autel, ou bien encore qu'ils placaient sur des crédences latérales. Selon Bocquillot et plusiers autres Liturgistes, it n'y aurait par aujour'dhui quatre siecles que les chandeliers sont devenus un décoration permanente de l'autel.—Anciennment il n'y avait sur l'autel ni livre ni cartons, ni croix. La table du sacrifice ne portait que les vases sacres.—*Liturgie Catholique par M. L. Abbe Migne.*

is a plain crucifix. The same author gives in the same work a series of illustrations respecting the investiture of Knights of the Bath, taken from an illuminated MS. of the latter part of the fifteenth century. In these are introduced altars with the crucifix and accompanying images of St. Mary and St. John, placed on the altar between two lights, with curtains on either side the altar.

The placing of more than two lights on an altar seems never to have been practised in the churches of this country. The Episcopal Injunctions in the thirteenth and two following centuries are confined to two, and I have not met with any ancient illumination in which

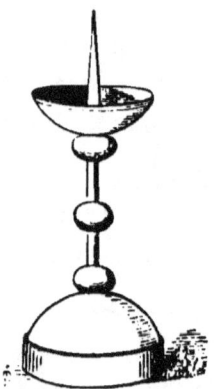

Pricket Candlestick, 12th Century, from a Crypt Painting, Canterbury Cathedral.

more than two are represented. In the fifteenth century, perhaps earlier, the high altar was inclosed on either side with curtains, suspended on rods of iron projecting from the walls.

The church of Clapton-in-Gordano, Somersetshire, still retains the ancient altar candlesticks of latten, of the fifteenth, or early part of the sixteenth, century, of one of which a representation is here given. Other articles of latten, base metal, or plate pertaining to the different services of the church may be enumerated;

Ancient Altar Candlestick of Latten, in the Church of Clapton-in-Gordano.

besides the chalice and paten, which I purpose treating of separately, there were the pix for the reservation of the Host, whence the monstrance was subsequently derived,[r] a pair of crewetts[s] of metal or glass, in which

[r] Dr. Harding, in his controversy with Bishop Jewell, mentions "the monstrauce or pixe" as if one and the same article.—*Defence of the Apology, &c.*, p. 343.

[s] In a rough draft of parish accounts, 7th and 8th Edward IV., of payments for the altar of St. Katherine, in Lapworth church, Warwickshire, occurs the following entry:—" It sol' pro ij. crwetts vi<sup>d</sup>." A few years ago a glass crewett was found concealed in this church; it is now in my possession.

were contained the wine and water preparatory to their admixture in the Eucharistic cup—a practice of the highest antiquity; a sacring bell, a pax table of silver[t] or other metal for the kiss of peace, which took place shortly before the Host was received in communion; this was sometimes of wood; a stoup or stok of metal, with a sprinkle, for holy water.[u] A censer or thurible,[x]

---

[t] A rich pax of silver, parcel gilt, is preserved at New College, Oxford. On it the crucifixion is represented with the attendant figures of St. Mary and St. John; and this was the most usual design. This pax is apparently of the latter half of the fifteenth century, and is figured in the 2nd Vol. of the *Journal of the Archæological Institute*.

[u] I have not met with any ancient example of this vessel in metal. From ancient illuminations it appears to have been a kind of open can, in shape resembling that of a flower pot, with a handle, and was carried by a clerk.

[x] The cover of an ancient thurible of latten was some years ago discovered in the church chest of Ashbury, Berkshire; the lower part is of a semi-globular, or domical form, from whence issues an embattled turret or lantern in the shape of a pentagon, which is finished by a quadrangular or four-sided spire; the sides, both of the lantern and spire, are partly of open-work, and round the globular part is inscribed *Gloria Tibi Domine*. This is now in my possession. I have also an ancient fenestral thurible of latten, of the fourteenth century, of which a representation is here given.

Fenestral thuribles were those where the sides were pierced in imitation of windows for the emission of the fumes of the frankincense. In an inventory, taken A.D. 1518, of church plate belonging to the Cathedral of St. Peter at York, are described, "Duo magna thuribula argentea deaurata, cum *fenestris*, &c. Duo thuribula argentea *plena fenestris*. Duo thuribula argentea unius sectæ *cum fenestris apertis* super conchas superiores." And in an inventory of jewels, ornaments, &c., of the Cathedral church of Lincoln, taken A.D. 1536, are mentioned, "Two pair of censers, silver and gilt, of bossed work, with four chains of silver, and every one of them a boss with two rings, having *six windows* and six pinacles." In or about the year 1851, on draining Wittlesea Mere a splendid silver thurible, weighing about fifty ounces, and a silver navicula, or ship for frankincense, were discovered. These were supposed to be of the fourteenth century, to have belonged to the Abbey of Ramsey, and to have been thrown into the Mere for concealment on the Suppression. Thuribles of latten or base metal are occasionally found from the twelfth century upwards. A great variety of

and a ship, (a vessel so called), to hold frankincense, preparatory to its being put into the thurible.

Censer or Thurible. 14th Century.

A CHRISTMATORY.[y] The MONSTRANCE, in which the Host was exhibited to the people, does not appear to

thuribles of diffcrent ages may be collected from ancient sepulchral monuments, the heads of the effigies on which are often supported by statuettes of angels or other figures waving thuribles, as on King John's Monument, Worcester Cathedral.

[y] Some 35 years ago a small christmatory box or coffer of latten was found concealed in St. Martin's church, Canterbury. A small ampulla of brass or latten, supposed to have been an ancient christmatory for the consecrated oil used in the sacrament of extreme unction, was, some years ago,

have been introduced into our churches before the fifteenth century. On the suppression of the monasteries we find it noticed in the inventories of church plate, then confiscated, as in that of the Priory of Ely, where it is called "a stonding monstral for the sacrament;" and in that of St. Augustine's Monastery, Canterbury, where it is described as "one monstrance silver gilt with four glasses."

In the twelfth and thirteenth centuries the pix for the reservation of the Host was a small circular box of metal often enamelled with Limoge work, with a conical lid surmounted with a cross.* A pix somewhat different to the above, but without its cover, of the metal called latten, was some years ago preserved in the church of Enstone, Oxfordshire; the body of this was of a semi-globular form, supported on an angular stem, with a knob in the middle, and in appearance not unlike a chalice.

Of the various vessels required for service at the altar, the most indispensable were the CHALICE and PATEN. Previous to the thirteenth century, for the reasons adduced, articles of church plate are exceedingly rare. We have, indeed, notices that church plate be-

---

discovered in the castle ditch, Pulford, Cheshire; this curious little relic was not more than two inches high, the body semi-globular or bulging in front, with a plain Greek cross engraved on it, and flattened at the back. At the neck were two bowed handles, by chains attached to which it appears to have hung suspended from the shoulders.

* I have a pix of this description in my possession of the thirteenth century, covered with enamel. A pix of a similar description, copper gilt and covered with enamel, said to be of the twelfth century, is engraved in the 1st Vol. of *Shaw's Dresses and Decorations of the Middle Ages*. I have met with other pixes of a similar type.

longing to the greater, that is conventual churches, even in Anglo-Saxon times, was exceedingly costly, and the materials of the precious metals, gold and silver. Of such was the gold pectoral cross, and the silver plate apparently the covering of a portable altar, *altare portabile*, found in the coffin of St. Cuthbert at Durham. These are articles perhaps of the earliest and rarest description we have existing in this country. In the year of our Lord 1070, William I. is said to have despoiled the English monasteries of gold and silver, not sparing even the chalices.[a] Changes took place in fashion in each succeeding age, and towards the close of the twelfth century, in or about the year 1193-4, most of the articles of church plate throughout the kingdom, more especially the chalices, were disposed of to raise the ransom from captivity of King Richard I., fixed at 180,000 marks.[b]

By the constitutions of William de Bleys, or Blois, Bishop of Worcester, A.D. 1229, two chalices were required for every church, one of silver to be used at mass, the other unconsecrated and made of tin, with which the priest was to be buried. It is from the chalices and patens found in the graves of ecclesiastics of priestly rank and upwards that we have the form of the ancient chalices of the thirteenth and fourteenth

---

[a] Anno Domini, MLXX. Rex Willielmus pessimo usus concilio, omnia Anglorum monasteria auro spolians et argento insatiabiliter appropriavit et ad majora Sanctae ecclesiae approbria calicibus et feretris non pepercit. —*Matt. Paris.*

[b] Unde calices angliae .... ad ejusdem regis liberationem cesserum in alienam possessionem. Unde Anglia ruina incurrit et damnum irrestaurabile.—*Matt. West.*

centuries; those found in the graves of bishops being generally of silver, those of inferior dignitaries, as priests, being of tin, latten, or other base metal, more or less corroded or broken. Sometimes the chalices deposited with the dead were of wax, as those with the monks of Durham. I have been unable to ascertain when the practice commenced of depositing a chalice, and in some instances also a paten on the breast of the body of a deceased prelate or priest, but it was undoubtedly a custom very generally observed in the thirteenth and fourteenth centuries; however, I have not met with chalices distinctively of a later period so found. Some chalices found with the remains of certain of the Archbishops of York, having been re-gilt, are, I believe, now used in that Cathedral at the celebration of the Holy Communion. In or about the year 1862, a silver-gilt paten, with a hand engraved in the centre—a not unusual device—was found in the stone coffin, in Worcester Cathedral, of Walter de Cantilupe, Bishop of Worcester, who died A.D. 1266. A chalice and paten of base metal was discovered some years ago in a grave in the churchyard of Sandford, Oxfordshire.

Of chalices of base metal deposited in the graves of priests I have two, or rather the fragments of two; the one found in the grave of a priest in Theddingworth churchyard, Leicestershire; the other, with a paten, found in the grave of a priest in Saccomb church, Hertfordshire. These are of the thirteenth or fourteenth centuries.

I have not met with any chalice, so early as the thirteenth or fourteenth century, retained in any of our churches. Some few, however, of the fifteenth century are still preserved. The chalice of the fifteenth century was in fashion different to those of an earlier period; the bowl was sometimes as small, but the stem was

Chalice and Paten, Sandford, Oxfordshire,
13th or 14th Century.

elongated, the knob angular-shaped and decorated, and the foot larger and more ornamented, and ofttimes sexagonal.

In Nettlecomb church, Somersetshire, are preserved an ancient chalice and paten of the middle of the fifteenth century. The bowl of the chalice is hemispherical, or nearly so, the foot sexagonal; on one of the curved sides of the stem, between the knob and the foot, is engraved the image of the Crucified; in the middle of the paten is engraved the head of our Saviour. Both the chalice and paten have the half-mark, a semifleur-de-lis, the leopard's head, and the letter B—a very early instance of the kind.[c]

In Trinity College, Oxford, are preserved an ancient

[c] Mr. Octavius Morgan, M.P., has expressed an opinion that these articles of church plate cannot be later than A.D. 1459.

chalice and paten, of the early part of the sixteenth century, A.D. 1527. Round the paten is engraved, "*Calicem salutaris accipiam et nomine Domini invocabor.*" In the centre of the paten is engraved a representation of the *Veronica,* or face of Christ. The bowl of the chalice is plain, but it appears to have the same inscription as that on the paten. This chalice is said to have belonged originally to the Abbey of St. Albans, Herts; and to have been given by Henry VIII. to Sir Thomas Pope, founder of Trinity College. Of the confiscation of church plate in the reigns of Henry VIII. and Edward VI., and of the change from the chalice to the Elizabethan communion cup, I purpose to treat in a future chapter.

Near the high altar we frequently find, in the south wall of the chancel, a series of stone seats, sometimes without, but generally beneath plain or enriched arched canopies, often supported by slender shafts, which serve to divide the seats. In most instances these seats are three in number, but they vary from one to five, and are the SEDILIA, or seats formerly appropriated during high mass to the use of the officiating priest and his attendant ministers, the deacon and sub-deacon, who retired thither during the chanting of the *Gloria in excelsis,* and some other parts of the service.[d] The sedilia sometimes preserve the same level, but generally they graduate or rise one above another; and that nearest the altar, being the highest, was occupied by

---

[d] Quo finito sacerdos cum suis ministris in sedibus ad hos paratis se recipiant et expectent usque ad orationem dicendam vel alio tempore usque ad *Gloria in excelsis.—MS. Rituale pen. Auc.*

the priest; the other two by the deacon and sub-deacon in succession.* We do not often meet with sedilia of so early an era as the twelfth century; there are, however, instances of such, as in the church of St. Mary, at Leicester, where is a fine Norman triple sedile, divided into graduating seats by double cylindrical shafts with sculptured capitals; and the recessed arches they support are enriched on the face with a profusion of the zig-zag moulding. In the south wall of the choir of Broadwater church, Sussex, is a stone bench beneath a large semicircular Norman arch, the face of which is enriched with the chevron or zig-zag moulding. In Avington church, Berkshire, is a stone beneath a plain segmental arch. Norman sedilia also occur in the churches of Earls Barton, Northamptonshire, and of Wellingore, Lincolnshire. From the commencement of the thirteenth century up to the Reformation sedilia became a common appendage to a church, and the styles are easily distinguished by their peculiar architectonic features. Some are without canopies, and are excessively plain. On the south side of the chancel of Minster Lovel church, Oxfordshire, is a stone bench without a canopy or division; and plain stone benches thus disposed are found in the chancel of Bloxham church,

* This arrangement was different to that directed by the rubrical orders of the Roman missals, on their revision after the Council of Trent, by which the celebrant was to be seated between the deacon and sub-deacon: "In missa item solemni celebrans medius inter diaconum et sub-diaconum sedere potest a cornu epistolæ juxta altare cum cantatur *Kyrie eleison, Gloria in excelsis*, et *Credo.*"—*Missale Romanum, Antverpiæ*, MDCXXXI.; *Rubricæ Generales*, &c. One of the queries published by Le Brun, whilst composing his liturgical work, was, "Si le prêtre s'assied au dessus du diacre et du soudiacre, ou au milieu d'eux."

Oxfordshire, and of Rowington church, Warwickshire. In Sedgeberrow church, Gloucestershire, are two sedilia without canopies; and in Standlake church, Oxfordshire, the sedilia, three in number, are without canopies or ornament. In Spratton church, Northamptonshire, is a stone bench for three persons, under a plain recessed pointed arch. In Prior's Hardwick church, Warwickshire, is a sedile for the priest, and below that, one double the size for the deacon and sub-deacon; both are under recessed arched canopies. Quadruple sedilia occur in the churches of Turvey and Luton, Bedfordshire; in the Mayor's chapel, Bristol; in Gloucester Cathedral; in the church of Stratford-upon-Avon, Warwickshire; and in Rothwell church, Northamptonshire; these are beneath canopies, and most of them are highly enriched. Quintuple sedilia sometimes occur, but are very rare; in the conventual church of Southwell, Nottinghamshire, are, however, five sedilia beneath ogee-headed canopies, richly ornamented. A single sedile for one person only is occasionally met with, but not often.

Eastward of the sedilia, in the same wall, is a *fenestella* or niche, sometimes plain, but often enriched with a crocketted ogee or pedimental hood moulding in front, over the arch, which is trefoiled or cinquefoiled in the head. This niche contains a hollow perforated basin or stone drain, called the PISCINA or LAVACRUM,*f* into which

---

*f* Prope altare collocatur Piscina seu Lavacrum in quo manus lavantur.—*Durandi Rat. de Ecclesia*, &c. In ancient church contracts the term *Lavatorie* was sometimes used for the Piscina, as in that for Catterick Church. In the Roman Missal subsequent to the Tridentine Council the word *Sacrarium* is used.

it appears that after the priest had washed his hands, which he was accustomed to do before the consecration of the elements, and again after the communion, the water was poured, as also that with which the chalice was rinsed. The usage of washing the hands before the communion is one of very high antiquity, and is expressly noticed in the Clementine Liturgy, and by St. Cyril in his mystical Catechesis;[g] we do not, however, find the piscina in our churches of an era earlier than the twelfth century, and even then it was of uncommon occurrence;[h] but in the thirteenth century the general introduction is observable. In Romsey church, Hampshire, is the shaft and basin (the latter cushion-shaped) of a curious Norman piscina; this is now lying loose, in a dilapidated state. In the south apsis of the same church is another Norman piscina, consisting of a quadrangular-shaped basin projecting from the south wall; and on the south side of the chancel of Avington church, Berkshire, is a plain Norman piscina within a simple semicircular arched recess. The churches of Kilpeck, Herefordshire; Keelby, Lincolnshire; and Bapchild, Kent, also contain Norman piscinæ. Those of all the various styles of later date are common; they exhibit, however, an interesting variety in design and

[g] At Alvechurch, Worcestershire, within the last few years, the custom prevailed of the priest washing his hands in the vestry before the administration of the sacrament, and napkins were brought to dry his hands. This custom also prevailed, as I am informed, in the church of St. George the Martyr, Queen's Square, London.

[h] The earliest reference I can find is that mentioned by Bede, with respect to the water in which the bones of King Oswald were washed, "ipsam que aquam in qua laverunt ossa in angulo sacrarii fuderunt."—*Bedæ Hist. Eccl. lib. iii. c. xi.*

ornamental detail. The drain of the piscina communicated with a perforated stone shaft, commonly enclosed in the wall, through which the water was lost in the earth; as in the case of the piscina with its shaft taken

Norman Piscina, Romsey Church, Hants.

out of the south wall of the chancel of the now destroyed church of Newnham Regis, Warwickshire. Sometimes a piscina was a subsequent addition to a structure of early date, as in the old and now demolished church of Stretton-upon-Dunsmore, Warwickshire, in the south wall of the Norman chancel of which a piscina of the latter part of the thirteenth century had been inserted.

The piscina is very common in churches even where the sedilia or stone seats are wanting, and not only in the chancel, but also in the south walls at the east end of the north and south aisles, and in mortuary chapels, as will be presently noticed; it appears, in short, to have been an indispensable appendage to an altar.

Sometimes, but rarely, the piscina is found in the east wall of the chancel, as in Ellesmere church, Salop, and occasionally in the east wall of a transept.

Sometimes the piscina is double, and contains two basins with drains, the one for receiving the water in which the hands had been washed, the other for the re-

Piscina, Newnham Regis, Warwickshire.

ception of the water with which the chalice was rinsed after the communion.[i] In Rothwell church, Northamptonshire, on the south side of the chancel, are the vestiges of a triple piscina; the fenestella has been destroyed, but the three basins with their drains remain. And in the ruins of Salley Abbey, Yorkshire, is a piscina with three basins.

[i] "Il y avoit pour cet effet en chaque piscine, comme on peut voir encore à une infinité d'autels, deux conduits, ou canaux, pour faire écouler l'eau, l'un pour recevoir l'eau qui avoit servi au lavement des mains, l'autre pour celle qui avoit servi au purification ou perfusion du chalice."—*De Vert, Explication des Cérémonies de l'Eglise, vol. iii. p.* 193.

What are called ground piscinæ, that is piscinæ level with the pavement, are found in the ruins of some of the ancient churches of the great Cistercian Abbeys in Yorkshire.

Across the fenestella, or niche which contains the piscina, a shelf of stone or wood may be frequently found: this was the CREDENCE,[k] or table on which the chalice, paten, ampullæ, and other things necessary for the celebration of mass were, before consecration, placed in a state of readiness on a clean linen cloth; and this originated from the πρόθεσις, or side table of preparation, used in the early church; a recurrence to which ancient and primitive custom by some of the divines of the Anglican Church, after the Reformation, occasioned great offence to be taken by the Puritan seceders. In some instances a side table of stone or wood was used for this purpose; and a fine credence table of stone, the sides of which are covered with panelled compartments, is still remaining on the south side of the choir, St. Cross church, near Winchester.[l]

The credence table, or shelf above the piscina, must not be confounded with the AMBRIE or LOCKER, a small

[k] In "*Le Parfaict Ecclesiastique, par M. Claude de la Croix*," (a curious work published A.D. 1666, and containing full instructions for the clergy of the Gallican Church, and an exposition of the rites and ceremonies,) amongst appendages to an altar is enumerated "une credance ou niche dans le mur a poser les burettes et le bassin," *p.* 536. And in another place, "au costé de l'Autel il y faut une petite niche à poser les burettes et le bassin, et y faire un trou en facon de piscine a fin que l'eau se perde en terre," *p.* 568.

[l] "In cornu Epistolæ . . . . ampullæ vitreæ vini et aquæ cum pelvicula et manutergio mundo in fenestella seu in parva mensa ad hæc præparata." —*Missale Romanum ex Decreto, &c.* 1631.

"Calix vero et alia necessaria præparentur in credentia cooperta linteo, antequam sacerdos veniat ad altare."—*Ibid.*

## PREVIOUS TO THE REFORMATION.

square and plain recess usually contained in the east or north wall, near the altar. In this the chalice, paten, and other articles pertaining to the altar were kept when not in use. The wooden doors formerly affixed to these ambries have for the most part either fallen into decay or been removed, but traces of the hinges may be frequently perceived; and a locker in the north wall of the chancel of Aston church, Northamptonshire, still retains the two-leaved wooden door. On the north side

Ambrie or Locker.

of the chancel of Floore church, Northamptonshire, is a very complete ambrie or locker with its original doors, ornamental hinges, and closing ring. Sometimes shelves are set across the lockers. In the east wall of Earls Barton church, Northamptonshire, is a large locker divided into two unequal parts by a stone shelf inserted in it; and in the north aisle of Salisbury Cathedral are two large triangular-headed lockers or ambries, each of which contains two shelves.

EASTER SEPULCHRES. Within the north wall of the chancel of many churches, near the altar a large arch like that of a sepulchral arch, more or less decorated, may be perceived; within this the HOLY SEPULCHRE, generally a wooden and moveable structure, was set up at Easter, when certain rites commemorative of the Burial and Resurrection of our Lord were anciently performed with great solemnity: and of these I purpose treating somewhat at length.

Bishop Jewel, in *The Apologie of the Church of England*, makes the observations following:—" As touching the multitude of vaine and superfluous ceremonies wee knowe that S. Augustine did grevousley complain of them in his own time: and therefore have wee cutte of a greate number of them because wee knowe, that mennes consciences were encumbred aboute them, and the Churches of God ouerladen with them. Neverthelesse wee keepe stil, and esteeme, not onely those ceremonies, whiche, we are sure, were delivered us from the Apostles; but somme others too besides, whiche we thought might be suffered without hurte to the Church of God; for that wee had a desire, that al thinges in the holy congregation mighte, as S. Paule commaundeth, be donne with comelinesse and in good order. But, as for al those thinges, whiche wee sawe were, either very superstitious, or utterly unprofitable, or noisome or mockeries, or contrarie to the Holy Scriptures, or els unseemely for sober and discreete people, these, I saie wee have utterly refused without al manner exception."

More than three centuries have elapsed since the

change in rites and ceremonies, which, apart from the doctrinal matters, distinguish the Church of England from the Church of Rome, yet there remain in our churches vestiges of antiquity, which have undergone more or less injury, and which carry us back to usages and customs formerly of common occurrence, but now little understood, and by few.

Amongst the ceremonies alluded to by Bishop Jewel as having been *cutte* off, were those of the adoration of the cross on Good Friday, and those commemorative of the Resurrection on Easter Day, connected as they were with those features of sculptured and architectural designs in our churches popularly known as Easter Sepulchres. These, however, were rather receptacles for the moveable sepulchres, which were of wood, than the actual sepulchres themselves.

It is on the origin of these sepulchres I now treat.

In that learned work of a celebrated divine of the Gallican Church, *Explication simple litterale et historique des Ceremonies de L'Englise*, written towards the close of the seventeenth century, par Don Claude de Vert, a more complete exposition of the origin of this custom is given than by any other writer I have met with. He states that it commenced at Jerusalem in the fourth century; that is to say, after the time that Saint Helen, mother of Constantine, had discovered our Saviour's cross, of which, says he, she left a part to be preserved by the Bishop of Jerusalem, and sent the rest to the Emperor, her son. The Bishop of Jerusalem exposed every year, on Good Friday, this part of the true cross to be adored; that is, to be saluted and reverenced by all the people. Thenceforth this ceremony of the exposition and adoration of the cross was communicated abroad to churches which possessed some morsel or small portion of this sacred wood (of which there were great numbers, about the midst of the fourth century, as reported by St. Cyril of Jerusalem), then by extension and imitation among the rest of the churches in the world; where, in default of some part of the true cross, common crosses were substituted, representing that of our Lord, which were exposed to the worship of the faithful; retaining always these words, *Ecce lignum crucis, in quo salus mundi perpendit.*ᵐ

The alleged discovery of the very cross on which our Lord suffered is said to have taken place A.D. 326. Eusebius, who died about A.D. 338, is silent about this discovery. St. Cyril of Jerusalem, who delivered his Catechetical Lectures about A.D. 347, speaks of the holy cross as discovered.ⁿ St. Ambrose (A.D. 395) and St.

---

ᵐ "Bois sacré de la Croix de Notre-Seigneur exposé dans les Eglises, le Vendredy Saint. Exposition qui a commencé à Jérusalem, dès le iv. siecle: c'est-à-dire, dès le temps que Sainte Hélene, mere de Constantin, eût découvert la Croix du Sauveur; dont elle laisse une partie en garde à l'Evêque de Jérusalem, pour la conserver, & envoya l'autre à l'Empereur son Fils. L'Evêque de Jérusalem exposoit tous les ans, le Vendredy-Saint, cette partie de la vraye Croix, pour être adorée c'est-à-dire, pour être saluée & révérée de tout le peuple. Et de là cette cérémonie de l'exposition & de l'adoration de la Croix se communiqua d'abord aux Eglises qui possedoient quelque morceau ou particule de ce Bois sacré (qui étoient déja en grand nombre, vers le milleu du iv. siecle, au rapport de S. Cyrille de Jérusalem, Catech., 4—10); puis par extension & imitation, dans tout le reste des Eglises du monde; où, au défaut de quelque partie de la vraye Croix, on substitua des Croix communes, Images de celle de Notre-Seigneur qui furent exposées de même au culte des Fideles; retenant toujours ces paroles, Ecce lignum crucis, in quo salus mundi perpendit."

ⁿ Sed magna planè tenet omnes admiratio; quid sil quòd Eusebius œdi-

Chrysostom (circa A.D. 394) speak of three crosses as discovered. The belief in the discovery of the cross was general at the close of the fourth century. The *Inventio Crucis* is commemorated in the Calendar prefixed to the Book of Common Prayer on the third of May.

The earliest allusion to this Rite of Adoration, as a stated ceremony to be observed, I have been able to meet with, is in the *Ordo Romanus*. This is said to have been compiled or composed by Gelasius, but subtracted from, and added to, and revised by Gregory the Great, at the close of the sixth century, about the year 597.

In this is the *Ordo in die Parasceves, ubi mos est salutiferam salutare crucem.* This Ordo ordains that at even the cross is prepared before the altar, a space intervening between it and the altar, carried by two acolytes. Then after certain texts of Scripture followed the anthem,—*Ecce lignum crucis, in quo salus mundi perpendit, venite: adoremus.*

Then comes the Pope alone, and adoring kisses the cross. Then the bishops, priests, deacons, and others in order, last of all the people. After sundry genu-

---

ficationem omnium harum basilcarum relegens, et de Helena prolixiorem mentionem faciens, de Cruce ub ea inventa ne verbum quidem; cum eam repertam esse sub Constantino, nulli sit dubium, sic scribente Cyrillo Episcopo Hierosolymitano ad Constantium Constantini ipsius filium Imperatorem: Tempore quidem (inquit) Deo dilectissimi, et beatæ memoriæ Constantini patris tui salutare Crucis lignum Hierosolymis repertum est, &c. Scimus tamen eundem Eusebium inventionis S. Crucis in Chronico meminisse, sed alio (ut videtur) tempore . . . . . . . . . . Cœterùn quòd Eusebii Chronicon a librariis admodun esse depravatum, &c.—*Baronii Annales*, sub anno 326.

flexions and prayers, follows the ancient hymn of the Latin Church commencing thus :—

"Crux fidelis inter omnes
Arbor una nobilis," &c. ;

this hymn having being concluded, and the cross saluted, it was deposited in its place.

Such in substance was the ancient rite of the Roman Church.

I do not find that Saint Isidore, of Seville, alludes to this ceremony. But Albinus Flaccus Alcuinus (who died A.D. 804), in his exposition of the Ordo Romanus, alludes, though more at length, to the salutation of the cross.

Amalarius, Archbishop of Treves, in the early part of the ninth century, (he died A.D. 837) in his work *De ecclesiasticis officiis*, treats " De adoratione sanctæ crucis," that the cross should be prepared before the altar, which cross all should salute and kiss.

Rabanus Maurus (who died A.D. 856), treating " De Parasceve," alludes to the salutation of the cross.

Durandus, who in the thirteenth century wrote that well-known work *Rationale divinorum officiorum*, treats also of the deposition and adoration of the cross, but his comments are full of allegorical and mystical meanings.

In a manuscript I possess of the fifteenth century, of the use of Sarum, are certain rubrical directions relative to this rite. The procession was to go through the west door to the place of the first station, on the north side of the church. Then the priest put off his chesible, and took the cross with his feet unshod and in his surplice, and deposited the cross in the sepulchre, then the host, but in a pix, in the same sepulchre.

The rubrical directions in a printed breviary according to the use of Sarum, a scarce but somewhat well-known work published by Merlin, at Paris, in the year 1556, are more lengthy than those in the manuscript I have alluded to.

*Martene De Antiquis Monachorum Ritibus*, published A.D. 1690, treats "De Resurrectione Dominica," as follows:—

"Post tertium . . . . quatuor fratres induant se, quorum unus alba indutus ac si ad aliud agendum ingrediatur, atque latenter sepulchri locum adeat; ibique manu tenens palmam quietus sedeat; dumque tercium percelebratur residui tres succedant, omnes quidem cappis induti, thuribula cum incenso manibus gestantes ac pedetentim ad similitudinem quœrentium quid, veniunt ad locum sepulchri; aguntur enim hæc ad imitationem Angeli sedentis in monumento, atque mulierum cum aromatibus venientium, ut ungerent corpus *Jesu:* Cum ergo ille residens tres velut erroneos ac aliquid querentes viderit sibi adproximare, incipiat mediocri voce dulcisone cantare: *Quem quœritis?* Quo decantato fine tenus; respondeant hi tres uno ore: *Jesum Nazarenum;* quibus ille *non est hic, surrexit sicut prœdixerat, ite nunciate quia surrexit a mortuis* . . . . . . . rursus ille residens velut revocans illos, dicat antiphonam *Venite et videte locum:* Hæc verò dicens surgat & erigat velum ostendantque eis locum cruce nudatum; sed tantum linteamina posita; quibus crux involute erat."

In the *Concilia Magnæ Brittanniæ et Hiberniæ*, better known perhaps as *Wilkins' Concilia*, I do not find in any of the Synodical or Provincial Constitutions relating to church furniture, any order for the sepulchre in the articles therein enumerated as essential to a church. It appears to have been regarded much in the light of organs to our churches, the gifts of individual benefactors, whilst the arches under which the Easter Sepulchres were placed, or the architectural and sculptured compositions within which they were deposited, and which at the present day are popularly, but erroneously, known as Easter Sepulchres, bore the same reference to the sepulchres as the organ-lofts or organ-chambers bore to the organ.

The ceremonies also were not the same in all churches, those observed on Easter Day being an addition to the more ancient rites observed on Good Friday. I have not as yet been able to trace the period when or about when the ceremonies on Easter Day commenced.

The earliest account of the sepulchre thus set up I have yet met with occurs in an inventory of church furniture, A.D. 1214, in which is mentioned *velum unum de serico supra sepulchrum.*

In a manuscript relating to Melford church, Suffolk, by Roger Martin, Esquire, of Melford Place, who lived at the time of the Reformation, the following particulars are given respecting the Easter Sepulchre:—

"In the quire, there was a fair painted frame of timber, to be set up about Maunday Thursday, with holes for a number of fair tapers to stand in before the sepulchre, and to be lighted in service time. Sometimes it was set overthwart the quire before the high altar, the sepulchre being alwaies placed, and finely garnished, at the north end of the high altar; between that and Mr.

Clopton's little chappel there, in a vacant place of the wall, I think upon a tomb of one of his ancestors, the said frame with the tapers was set near to the steps going up to the said altar."

The tomb thus noticed is a rich canopied tomb between the quire and the Clopton chapel, and is the tomb of John Clopton, Esquire, of Kentwell Hall, Sheriff of the County of Suffolk, in 1451, and who died in 1497.

In an inventory of plate and goods belonging to Melford church, the following mem. appeared :—"Mem., April 6, 1541.—There was given to the Church of Melford, two stained cloths, whereof the one hangeth towards Mr. Martin's ile, and the other to be used about the sepulchre at Easter time."

On the north side of the chancel of Stanwell church, Middlesex, is a high tomb, over which is a canopy with an obtuse arch, ornamented with quatrefoils. Beneath the arch were placed upright in the wall, brass plates with the effigies of the deceased and his wife, all long ago removed. This is the monument of Thomas Windsor, who died A.D. 1486. By his will made in 1479, after directing that his body should be buried "on the north side of the quer of the church of our Lady of Stanwell, afor the ymage of our Lady, wher the sepulture of our Lord stondith;" he adds "I will that there be made a playne tombe of marble, of a competent height, to th' entent that yt may ber the blessed body of our Lord, and the sepulture at the time of Estre to stand upon the same."

Bloomfield and Parkin, the historians of Norfolk, in the account of Raineham, say, speaking of Eleanore, wife of Roger Townsend, Justice of the Common Pleas, temp., Henry VIII., and which Eleanore survived the judge her husband, and died A.D. 1500: she by will dated November 9, 1499, orders her body to be buried by the high altar, before our blessed Lady in the chancel of Rainham St. Mary, and a new tomb to be made for her husband's and her bones; upon which tomb to be cunningly graven a sepulchre for Easter Day.

In the description of Raineham church, it is stated: Against the north-east part of the chancel is a very fair tomb, but without any inscription, erected to the memory of Sir Roger Townsend, the judge; agreeable to the will of the Lady Elianore his wife, as is above mentioned, with a canopy, &c., of stone work.

In the constitutions of the office of first Deacon of Trinity church, Coventry, A.D. 1452, is this—"Also he schall wache ye sepulcur on Aster evyn tyll ye resurrecion be don."

The second deacon's constitutions also provide that "he shal wache ye sepulcur on gode fryday all nyght."

A sepulchre belonging to St. Mary Redcliffe church, at Bristol, is in a document thus described:—

"Item, that Maister Canyne, hath delivered this 4th day of July, in the year of our Lord 1470, to Maister Nicholas Petters, Vicar of St. Mary Redcliffe, Moses Conterin, Philip Barthelmew, Procurators of St. Mary Redcliffe aforesaid, a new sepulchre gilt with golde, and a civer thereto. Item, an image of God Almighty, rising out of the same sepulchre with all the ordinance that longeth thereto, that is to say, a lathe made of timber and the iron work thereto. Item, thereto longeth

Heaven made of timber and stayned clothes. Item, Hell made of timber thereto, with Divils to the number of 13. Item, 4 Knights armed, keeping the sepulchre with their weapons in their hands; that is to say 2 axes, and 2 spears, with 2 paves (*i.e.* shields). Item, 4 payr of Angels wings, for 4 Angels made of timber, and well painted. Item, the Fadre, the Croune, and Visage, the ball with a cross upon it, well gilt with fine gould. Item, the Holy Ghost coming out of Heaven into the sepulchre. Item, longeth to the 4 Angels, 4 Chevelures " (*i.e.* perukes).

I can imagine a gorgeous piece of carved and joiner's work mixed with architectural detail, in the above-described composition. In that curious, but scarce and valuable little work, *The Ancient Rites and Monuments of the Monastical and Cathedral Church of Durham,* collected out of ancient manuscripts about the time of the Suppression, and published by J. D. (Davies), of Kidwelly, in 1672, a popular account is given of the rites and ceremonies used in that church, on Good Friday and Easter Day. This is said by Mr. Raine, who edited it for the Surtees Society a few years ago, to have been written in 1593. It may have been written by some one who had taken a part in the ceremonies, and was well conversant with them. He describes them as follows:—

"*The Passion.*—Within the Abbey church of Durham, upon Good Fryday, there was marvellous solemn service, in which service time, after the Passion was sung, two of the ancient monks took a goodly large crucifix all of gold, of the picture of our Saviour Christ nayled upon the cross; laying it upon a velvet cushion, having St. Cuthbert's arms upon it all embroider'd with gold, bringing it betwixt them upon the cushion to the lowest greeses, or steps in the quire, and there betwixt them did hold the said picture of our Saviour, sitting on either side of it. And then one of the said monks did rise, and went a pretty space from it, and setting himself upon his knees with his shoes put off, very reverently he crept upon his knees unto the said cross, and most reverently did kiss it; and after him the other monk did so likewise, and then they sate down on either side of the said cross, holding it betwixt them. Afterward the prior came forth of his stall, and did sit him down upon his knees with his shoes off in like sort, and did creep also unto the said cross, and all the monks after him, one after another in the same manner and order; in the meantime the whole quire singing a hymn.

"The service being ended, the said two monks carried the cross to the sepulchre with great reverence (which sepulchre was set up in the morning on the north side of the quire, nigh unto the high altar, before the service time), and there did lay it within the said sepulchre with great devotion, with another picture of our Saviour Christ, in whose Breast they did inclose with great reverence the most holy and blessed Sacrament of the Altar, censing and praying unto it upon their knees a great space; and setting two lighted tapers before it, which did burn till Easter Day in the morning, at which time it was taken forth."

"*The Resurrection.*—There was in the Abbey church

of Durham, very solemn service upon Easter Day, betwixt three and four of the clock in the morning, in honour of the Resurrection; where two of the eldest monks of the quire came to the sepulchre, set up upon Good Fryday, after the Passion, all covered with red velvet and embroider'd with gold, and did then cense it, either of the monks with a pair of silver censers, sitting on their knees before the sepulchre. Then they both rising came to the sepulchre, out of which with great reverence they took a marvellous beautiful image of our Saviour, representing the Resurrection; with a cross in his hand, in the breast whereof was enclosed, in most bright chrystal, the holy Sacrament of the Altar; through which chrystal the Blessed Host was conspicuous to the beholders. Then after the elevation of the said picture carried by the said two monks, upon a fair velvet cushion all embroider'd, singing the anthems of *Christus Resurgens*, they brought it to the high altar, setting it on the midst thereof, and the two monks kneeling before the altar, and censing it all the time that the rest of the whole quire were singing the foresaid antheme of *Christus Resurgens*: which antheme being ended, the two monks took up the cushion and picture from the altar, supporting it betwixt them, and proceeding in procession from the high altar to the south quire door, where there were four ancient gentlemen belonging to the Prior appointed to attend their coming, holding a most rich canopy of purple velvet, tassell'd round about with red silk, and a goodly gold fringe; and at every corner of the canopy did stand one of these ancient gentlemen, to bear it over the said images, with the Holy Sacrament carried by the two monks round about the church, the whole quire waiting upon it with goodly torches, and great store of other lights; all singing, rejoycing, and praying to God most devoutly, till they came to the high altar again; upon which they placed the said images, there to remain till Ascension Day."

In articles about religion set out by the Convocation, and published by the King's authority, A.D. 1536, one is *Of Rites and Ceremonies.* "As concerning the rites and ceremonies of Christ's Church—creeping to the cross, and humbling ourselves to Christ on Good Friday, and offering there unto Christ before the same, and kissing of it in memory of our redemption by Christ made upon the cross; setting up the sepulchre of Christ, whose body after His death was buried; . . . and all other like laudable customs, rites, and ceremonies be not to be contemned and cast away, but to be used and continued as things good and laudable, to put us in remembrance of those spiritual things that they do signify, not suffering them to be forgotten, or to be put in oblivion, but renewing them in their memories from time to time; but none of these ceremonies have power to remit sin, but only to stir and lift up our minds unto God by whom only our sins be forgiven."

About the year 1543, the rites and ceremonies of the Church were, in this country, brought under review, and a *Rationale* drawn up to explain the meaning and justify the usage. In this the rites of the Easter Sepulchre are stated and expounded as follows:—

"Upon Good Friday is renewed yearly the remembrance of the blessed passion: wherefore that day amongst other godly ceremonies to be continued, is the creeping to the cross, where we humble ourselves to Christ before the same: offering unto Him and kissing of the cross in memory of our redemption by Christ upon the cross.

"And that day is prepared and well-adorned the sepulchre in remembrance of His sepulchre, which was prophesied by the prophet Esaias to be glorious, wherein is laid the image of the cross, and the most blessed Sacrament, to signify that there was buried no corpse or body that could be purified or corrupted, but the pure and undefiled body of Christ, without spot or sin, which was never separated from the Godhead. And therefore, as David expresseth in the fifteenth Psalm, it could not see corruption, nor death could not detain or hold Him. but He should rise again to our great hope and comfort; and therefore the Church adorns it with lights, to express the great joy they have of that glorious triumph over death, the devil, and hell.

"Upon Easter Day in the morning, the ceremonies of His resurrection are very laudable, to put us in remembrance of Christ's resurrection, which is the cause of our justification. And that as Christ being our head, was the first among the dead which rose never to die again; so all Christian men being His members, do conceive thereby to rise from death of sin to godly conversation in this life; and finally, at the day of judgment, when the bodies and flesh of all mankind shall by the operations of God be raised again, to rise with Him to everlasting glory."

In 1559, Thomas Naogeorgus, published in Latin verse, at Basil, the *Regnum Papisticum*, a satirical work, in a great measure deirsive of the rites and ceremonies of the Church of Rome. This was "Englyshed," or rendered into English verse by Barnabe Googe in 1570, whose version I give—

*Good Friday.—*
"Two Priestes the next day following, upon their shoulders beare,
The Image of the Crucifix, about the altar nere:
. . . . . . . . . . . . . .
Another Image doe they get, like one but newly deade,
With legges stretcht out at length and handes, upon his body spreade:
And him with pompe and sacred song, they beare unto his graue,
His bodie all being wrapt in lawne and silkes and sarcenet braue,
. . . . . . . . . . . . . .
And least in graue he should remaine, without some companie,
The singing bread is layde with him, for more idolatrie:
The Priest the Image worships first, as fallest to his turne,
And frankensence and sweete perfumes, before the bread doth burne:
With tapers all the people come, and at the barriars stay,
Where downe upon their knees they fall, and night and day they pray:
And violets and every kinde of flowres about the graue
They straw, and bring in all their giftes, and presents that they have.
. . . . . . . . . . . . . .

*Easter-day.—*
At midnight then with carefull minde, they up to mattens ries,
The Clarke doth come, and after him, the Priest with staring eies:
The Image and the breade from out the graue (a worthie sight)
They take, and Angels two they place in vesture white,
. . . . . . . . . . . . . .
In some place solemne sightes and showes, and pageants fayre are play'd,
With sundrie sortes of maskers braue, in straunge attire arayd,
As where the Maries three doe meete, the sepulchre to see,
And John with Peter swiftly runnes, before him there to bee."

The Latin version of Naogeorgus is as follows:—

*Parasceve.* "Luce Sacerdotes duo circum altare sequenti
Idolum portant humeris Christi in cruce fixi,
. . . . . . . . . . . . .
. . . . . . . . . . . . .

Assumant aliam statuam pro more iacentem
Defuncti nuper, porrectis cruribus aptè,
Atq; decussatim compostis pectore palmis,
Et pompa cantuq; pio ad factum ante sepulchrū
Portant, sericeis tectam membra omnia peplis
. . . . . . . . . . . . .
Ne iaceat verò, inq; sepulchro sola colatur,
Mysticus adfertur quoq; et unà clauditur intus
Panis, ut impietas crescat cultusq; prophanus.
Sacrificus supplex statuam veneratur inertem
Primus, et Asyrios pani succendit odores.
Multa statim populus cōportat lumina circum,
Cancellisq; hœret, noctemq; diemq; precatur
Curuatis genibus, uiolisq; et flore sepulchrum
Omnigeno exornat, suaq; affert munera largè.
. . . . . . . . . . . . .

*Pascha.*  Nocte dein media consueta ad cantica surgunt
Prœvenit œdituus studio rasusq; sacerdos,
Et statuam é clauso tollunt, paṇemq; sepulchro:
Angelicasq; indunt statuas, peplа rara tenentes
. . . . . . . . . . . . .
Est ubi continuò ludi et spectacula dantur
Ut tres conveniant Mariæ, visantq; sepulchrum,
Cumq; Petro currat uelox Zebedeia proles."

In *The Beehiue of the Romish Churche*, translated out of Dutch into English, by George Gilpin the elder, newly imprinted 1580, a satirical production, this custom is thus adverted to—"And besides she (the Church of Rome), hath more ordained and charged that wee should upon the Good Friday after Maundy Thursday, devoutly and sadly creeping along the grounde upon our bare knees, worship the cross, and there bestowe a good fat offering and liberal almes to the benefit and maintenance of the poore Priests."

Now as to the disuse of these ceremonies.

In the royal injunctions issued in the first year of the reign of Edward VI. (A.D. 1547), I find no allusion to the ceremonies on Good Friday or Easter Day.

In a royal proclamation against those that do innovate, alter, or leave down any rite or ceremony, in the church of their private authority, set forth the 6th day of February, in the second year of the King's Majesty's most gracious reign—it was provided that for not creeping to the cross, no man hereafter be imprisoned or otherwise punished.

The custom was now left as a matter of indifference to be observed or discontinued according to the will of each individual. In some of the writings of the early reformers allusions are made to these rites, which are animadverted upon and decried.

In the visitation articles of Archbishop Cranmer, in the second year of the reign of Edward VI., one put to the clergy is, whether they had upon Good Friday last past the sepulchres with their lights having the sacrament therein.

In certain articles to be followed and observed according to the King's Majesty's injunctions and proceedings, to which articles I find no date assigned, the ninth enjoins amongst other rites, that no man maintain sepulchres, creeping to the cross.

In a letter with articles sent from the Queen's Majesty (*i.e.* Queen Mary) to the Bishop of London, in March, 1553, the thirteenth, is "that the laudable and honest ceremonies which were wont to be used, frequented, and observed in the church, be also hereafter frequented, used, and observed."

In injunctions issued by Cardinal Pole, in the Diocese of Gloucester, (in what year I know not), one appears touching the laity, "that all parishioners shall obediently use all the godlie ceremonies of the church as (amongst others therein enumerated) creeping to the cross."

Early in the reign of Elizabeth the rites on Good Friday and Easter Day, formerly observed, seem to have been generally discontinued.

Many of the ancient timber sepulchres appear to have been sold or broken up, in the reign of Edward VI., restored or remade in the reign of Mary, and finally again excluded from our churches early in the reign of Elizabeth. Ancient churchwardens' accounts in the middle of the sixteenth century, present items relative to these facts, as at an earlier period they did of payments for watching the sepulchre, and for the lights burning before it, whilst at the same time the oblations made at the sepulchre, and called "Creeping Silver," from the creeping to the cross, are also accounted for.

In that valuable work, *English Church Furniture*, ornaments and decorations at the period of the Reformation, as exhibited in a list of the goods destroyed in certain Lincolnshire churches, in the year 1566, most ably edited by Mr. Peacock, many of the timber sepulchres are accounted for, and some of the various uses to which they were put. A few extracts will not be out of place.

*Asbye iuxta Sleford.*—Itm., ōr sepulcre broken and burned a. o. ij. Elizabethe.

*Bastone.*—Itm., one sepulcre broken and defaced.

*Belton, in the Isle of Axholme.*—Itm., a sepulker with little Jack broken in pieces one year ago, but little Jack was broken in peces this yeare by the said churchwardens.

Little Jack, as it was irreverently called, was the box or pix in which the host or sacramental wafer was enclosed, and placed within the sepulchre.

*Birton.*—Itm., the sepulcre was burnte in melting lead for to mend or churche.

*Blyton.*—Itm., a sepulker of wainscot, taken from the church by the Vicar, and remaineth in his house as wee suppose.

*Castlebyth.*—Itm., one sepulcre wch we have made a coion (communion) table of.

*Croxton.*—Itm., a sepulker whearof is made a shelf for to set dishes on.

*Denton.*—Itm., one sepulcre, sold to John Orson, and he haith made a presse thereof to laie clothes therein.

*Durrington.*—Sepulcre was broken and sold to Willim Storre, and Robert Cappe, who have made a henne penne of it.

I need not instance more examples. I have only met with one wooden coffer I believe to have served as the Easter Sepulchre. There may be others still existing in country villages, used for domestic purposes, but as yet unheeded and unknown.

What appears to have been the moveable Easter Sepulchre formerly belonging, as I think, to Kilsby church, Northamptonshire, and now in the possession of a gentleman in Warwickshire, consists of a wooden coffer three feet nine inches in length, one foot three inches in width, and one foot nine inches in height, exclusive of modern supports. The cover is comparatively modern. The back, which was placed against the north wall of the chancel, is plain, but the ends and front have five square panels carved in relief, one at each end, and three in front. Each panel is about thirteen inches by eleven. These panels have each a group of figures. That at the east end, presuming this coffer to have been placed against the north wall of the chancel, represents our Lord clad in a long tunic—the coat without a seam—with His hands bound, standing before Pilate, who is represented clad in regal attire— *viz.*, the tunic and mantle with the hood worn over the shoulders and in front of the breast, with a close-fitting cap on his head. In the group of five figures of which this carving consists are two soldiers, one of which, with a figure in ordinary civil costume, is in the background, and hardly distinct enough to describe. The other soldier appears in the armour of a knight of the fourteenth century. He wears on his head a basinet, to which is attached a camail, or tippet of mail, covering the breast and neck, epaulieres protect the shoulders, and a long sleeveless surcoat is worn over the body armour. On the panel at the west end of the coffer, our Lord is represented clad in a long tunic, with a nimbus about His head, bearing His cross. The other figures in this group consist of a female, probably intended to represent St. Mary, and a soldier, the details of whose armour are not very clearly developed.

In front of this coffer are three groups; the westernmost appears to represent the deposition from the cross: this is partly destroyed. On one side is represented a man clad in a close-fitting jack, or short vest, with a bawdrick or broad girdle about the loins, holding in his hand a hammer; in the background appears a ladder.

Another figure, with a hammer or pincers, appears in the other side of the body of our Lord. Two other figures appear on the right of the cross, probably representing the Blessed Virgin and St. John.

The middle panel in front represents the resurrection of our Lord. He appears rising from the sepulchre, with the conventional cross in His left hand; His right hand, with two of the fingers erect, being upheld in the act of benediction. Four soldiers complete this group: of these, two appear in the background, standing behind the sepulchre; their lower limbs are consequently hid. They are represented in the armour of knights: basinets, with camails of mail, protect the heads, necks, and breasts; the shoulders, arms, and hands are defended by epaulieres brassarts, vambraces and gauntlets of plate; and their bodies are covered with close-fitting jupons worn over the plastron de fer; one holds a glavè, the other a pole-axe. Outside and in front of the tomb are grouped two soldiers in a sleeping attitude, in like armour as the former; one wears a dagger, or anelace, on his right side.

The last and easternmost panel in front represents our Saviour in the garden, after His resurrection, appearing to Mary Magdalen. The upper part of His body is nude, but the loins are covered with loose drapery.

This is the only moveable Easter Sepulchre of wood (for such I believe it to be) I have met with, and therefore I have somewhat described it at length. The carvings are not unlike those sculptures in relief in alabaster called "tables," so frequent in our churches in the fifteenth century, and some of which we occasionally meet with. From the hood which Pilate is represented as wearing, I attribute this carved sepulchre to the reign of Richard the Second, or the last twenty years of the fourteenth century.

And lastly, as to the fixed and permanent receptacles of masonry, of architectural and sculptured design, some consisting of a simple arch' only beneath which the moveable Easter Sepulchre was placed; others of more elaborate work to be found in some, but not the greater part of our churches; but in all instances, I think, constructed within, or attached to the north wall of the chancel.

I cannot find the introduction of these receptacles to have taken place earlier than the thirteenth century. There may be some of a prior period, but I have not yet met with such. I shall confine myself to describing a few instances.

At Cubbington church, Warwickshire, the receptacle consists of a low platform in the north wall of the chancel, beneath a plain pointed elliptical-shaped arch, devoid of sculpture or ornament. This appears to be of the thirteenth century.

In the north wall of the chancel of Long Itchington church, in the same county, is an arched recess at some height from the pavement, surmounted by a pedimental canopy. This appears also to be of the thirteenth century.

Within the north wall of the chancel of Garthorpe church, Leicestershire, is a recess surmounted by a plain pointed elliptic arch, but without any appearance of a tomb beneath; this appears to have been the receptacle for the Easter Sepulchre. Above this arch is a recess with a plain pedimental canopy; this appears to have served for the deposition of the pix containing the host. These two several receptacles, I think, are of the early part of the fourteenth century.

In the north wall of the chancel of Bilton church, Warwickshire, the receptacle for the Easter Sepulchre appears in a recess surmounted by an elegant and enriched ogee-shaped arch, elaborately moulded and crocketted, terminating with a finial. The architectural design of this may be attributed to about the middle of the fourteenth century.

In the north wall of the chancel of Stanton St. John, Oxfordshire, is a recess surmounted by an ogee-shaped arch, richly crocketted and cusped within, and presenting features very similar to those of the recess in Bilton church, and apparently of the same period.

The most interesting examples, however, of receptacles for Easter Sepulchres, wherein both architectural features of a high degree of merit and sculptured accessories skilfully worked out are combined, are to be met with in the churches of Hawton, and St. Peter, Sibthorp, Nottinghamshire; Heckington, Navenby, and Lincoln Cathedral, Lincolnshire; Patrington, Yorkshire; Northwold, Norfolk; and Holcombe Burnell, Devonshire.

That of St. Peter, Sibthorp, very recently discovered on the removal of a monument, contains above the arched recess for the sepulchre a niche above, apparently for the reception of the host.

That at Heckington is a fine and rich specimen of

architectural composition of the fourteenth century. In front, under four rich pedimental canopies, are represented four soldiers; above is a recess beneath a triangular arch; above this arch is the resurrection of our Lord, with two angels beneath. On either side of the arch are angels and figures of the three Maries. The whole design and composition is' worthy of attentive study.

In the chancel of Hawton church, Nottinghamshire, the receptacle for the Easter Sepulchre in general design resembles that at Heckington, though many of the features are different. The four soldiers, represented as watching the sepulchre, have, as defensive armour, basinets, hawberks of chain mail, with sleeveless surcoats over, and heater-shaped shields. On one side of this receptacle is a small recess, apparently for the deposition of the pix containing the host. Close to this receptacle, and of coeval date, is the fine monument and recumbent effigy of Robert de Compton, lord of the manor of Hawton in the reign of Edward III., by whom this beautiful receptacle for the sepulchre and his own monument, harmoniously grouped together, and presenting, both in general design and architectural and sculptured detail, the most chaste and elegant composition of the fourteenth century, appear to have been constructed.

The receptacle for the Easter Sepulchre in Lincoln Cathedral is very fine and interesting, and a rich specimen of architectural design of the fourteenth century. The composition consists of three trefoil-headed arches with pedimental canopies above, richly crocketted and finialled. In the three divisions at the base are represented, in reclining positions, three soldiers in hawberks and sleeveless surcoats, with heater-shaped shields.

At Patrington church, Yorkshire, the receptacle for the sepulchre does not appear of the same age as those at Hawton, Heckington, and Lincoln, but is probably a work of the early part of the fifteenth century. It represents the watching of three soldiers beneath three ogee-shaped canopies, and above is a representation of our Saviour emerging from the sepulchre, the lid of which is being raised by two angels.

A recess in the north wall of the chancel of East Kirby church, Lincolnshire, exhibits architectural features of the fourteenth century, but the only sculptured accessories are the bustos of three women, apparently representing the three Maries; but this receptacle has one peculiar feature belonging to it which I have not met with elsewhere, for, partly projecting from the ledge on which the Easter Sepulchre was placed, is a stone basin, without perforation, for the reception, I have no doubt, of the oblations it was customary to offer, in ancient accounts, as I have before observed, called "*Creeping silver.*"

At Northwold church, Norfolk, is a sculptured design for the reception of the sepulchre, in which are represented four soldiers watching; but this appears to be of the fifteenth century, and the sculpture is much defaced.

In the north wall of the chancel of Holcombe Burnell church, Devonshire, near the altar, is some sculpture in alto relievo, representing the resurrection and soldiers watching. This I have not yet seen, and therefore cannot speak of it more particularly.

In some alterations made in the chancel of Withybrook church, Warwickshire, in the year 1848, the remains of the receptacle for the Easter Sepulchre were discovered. These consisted of an arch three feet three inches across, the lower part panelled, and the soffit of the arch diapered; the ground was coloured azure, with white stars interspersed. At the back of the arch the ground was painted red, with white and blue leaves. At the back were sculptured angels and two armed figures of soldiers; these were much mutilated, but from fragments of taces attached to the breast-plate of one of them, this receptacle had apparently been constructed in the reign of Henry VI., or about the middle of the fifteenth century. Round the body armour of the other soldier was a bawdric.

The receptacles I have described present, one or more of them, the recess within which the moveable sepulchre was set, the recess for the pix containing the host, the basin for the creeping silver or offertory, and the group of soldiers for the watch.

In an inventory of church goods belonging to St. Margaret's church, Southwark, A.D. 1485, the following are enumerated:—

"Item ij blew Cortyns (to) draw afore the sepulture."

"Item a lytyll Cortyn of grene sylke for the hede of the scpulture."

"Item iij Cortyns of lavnde to draw afore the sepulture on the ester halydays."

"Item iij steyned Clothys with the Passyon and the

Resureccyon to hangg about the sepulture on good fryday."

"Item vi angelles of tre (wood) gylt with a tombe to stonde in the sepulture at Ester."

"Item iiij long crests (cressetts) & iiij short for to sett the lyghtes aboote the sepulture on good fryday, peynted rede with yrons to the same."

CONFESSIONALS. We nowhere find in the churches of this country, previous to the Reformation, any structures or articles of church furniture such as those we meet with in Continental churches, known as confessional boxes; and even these latter are hardly of earlier date than the seventeenth century. Leo the Thracian, Emperor of the East, is said, circa A.D. 458, to have recommended private confession of sins to a priest, in preference to public confession. Compulsory auricular confession was established by or at the time of the fourth Lateran Council, held A.D. 1215. By the Council of Durham, held A.D. 1217, the confessions of women were to be heard without the veil, and openly, *in propatulo*, as far as outward appearance was concerned, but still not so as to be heard by any. By the Synod of Exeter, A.D. 1287, women were to confess openly, and without the veil; not that they should be heard by the public, but seen.°

By the Constitutions of Walter Reynold, Archbishop of Canterbury, A.D. 1322, the priest about to hear confessions was to choose some common place in the church where he might be seen of all indifferently, and was

* Mulieres in aperto et extra velum confiteantur non ut audiri valeant sed videri.

not allowed to exercise that rite, especially as regarded women, in obscure places, except from urgent necessity, or from the infirmity of penitents.[p]

At a late period, however, we meet with notices of the shriving stool and pew.

In a list of goods in Great St. Mary's church, Cambridge, taken A.D. 1505, is the following:—"It' vj ycrnes perteynyng to the shryvyng stole for Lenton." And in the accounts of St. Michael's church, Cornhill, London, we have one in 1548, as follows:—"Payd to the joyntr for takynge down the shryvyng pew and making another pew in the same place."

A small erection of stone inside of Tanfield church, Yorkshire, may have been intended for a confessional, but I have met with no similar structure. It is entered from behind a pier on the north; it abuts against a wall on the west; on the east it has three small lights, one above the two; and on the west two small lights.

Thus far regarding confessions and the places for making the same, as heard by the parish priests; but in the thirteenth century, the secular clergy were interfered with by the preaching friars of the Dominican and Franciscan Orders, to whom permission was granted to hear confessions, and the precise places where they were accustomed so to do was subsequently and distinctly pointed out.

[p] Item sacerdos ad audiendum confessiones communem sibi eligat locum, ubi communiter ab omnibus videri poterit in ecclesia; et in locis absconditis non recipiat sacerdos alicujus, et maxime mulieris, confessionem, nisi pro maxima necessitate, aut infirmitate pœnitentis.—*Wilkins' Concilia.*

The Order of Dominicans, *Fratres prædicatores*, was established A.D. 1216. They are said to have come into England about A.D. 1221. The Franciscan Order of Mendicants was established A.D. 1223, shortly after which, about A.D. 1224, they are said to have come into England. Peculiar privileges were granted them. Pope Clement IV., A.D. 1265, empowered the Friars Minor (Franciscans) to preach, hear confessions, and to give absolution, and enjoin penance, without the assent of the secular clergy, *sine sacerdotum parochialium licentia.*[q]

In the latter part of the thirteenth century, A.D. 1287, the Archbishop of Canterbury granted license to the Friars Minor to hear confessions and give absolution without the assent of the parochial clergy.[r]

In the year 1278, Giffard, Archbishop of York, wrote to the Prioresses of Hampole, Appelton, Syningthwaite, and the rest of the houses of the Cistercian Order in the Diocese of York, that they should admit the Minorites and Preaching Friars for their Confessors (who, he says, "*in ecclesia Dei fulgent velut splendor firmamenti,*") notwithstanding the inhibition those nunneries had received from the Abbot of the Cistercian Order, who, he adds, "*in vos nullam habeat jurisdictionem ordinariam vel etiam delegatam.*"[s]

A.D. 1300, we find the Archbishop of Canterbury

[q] *Waddingi Annales, Minorum Tome ii. Regestum Pontificium, p.* 101.
[r] A.D. 1287, Archiepiscopi Cantuar litera conservatoria pro fratribus minoribus, quod possunt audire confessiones, et absolvere omnes fideles indistincte, irrequisitio consilio et assensu absque licentia paroch: presbyter.—*Wilkins' Concilia.*
[s] *Monasticon, vol.* 5, *p.* 99.

delivering his mandate for the admission of the Friars Minor to hear confession.'

The "asoiling" of the friars was subsequently the subject of invective against them by the medieval poets, as in the vision of Piers Ploughman, written circa A.D. 1362:—

> " Now that thow come to good confesse the to some frere
> He shal asoile the thus sone how thow evere wynne hit.
> . . . . . . . . . . . . . .
> And flittynge fond ich the frere that me confessede
> And saide he myghte me nat asoile but ich sulver hadde
> To restitue resonabliche, for al unryghtful wynnyng."

The peculiar LOW SIDE WINDOW, common in some districts, especially in churches erected in the thirteenth and fourteenth centuries, may, I think, clearly be connected with the practice of asoiling by the friars. This window is generally found in the south wall of the chancel, near the south-west angle, but sometimes on the opposite side, and occasionally even in one of the aisles,

---

' Mandatum Archiepiscopi Cantuar. de admissione fratrum minorum ad audiend confessiones.

Robertus &c., magistro Martino commissario Cantuar. salutem. Quia fratribus minoribus inferius nominatis, et vobis, post licentiam et gratiam nostram, a provinciali ministro ipsius ordinis ad subscripta petitam et obtentam, juxta constitutionem novellam domini papae ad id specialiter praesentatis; *viz.* fratribus Radulpho de Wodeheye, Johanni de Kenelden, Johanni de Bedewynde, Willelmo de Dele, Rogero Malemeyns, et Thomae de Malmesbur. licentiam in forma constitutionis ejusdem nuper dedimus specialem, ut confessiones quorum cunque utriusque sexus nostrae Cantuar dioce eisdem fratribus vel eorum alicui sua peccata confiteri volentium, audiaut et eis imponant poenitentias salutares, ae beneficium absolutionis impendant, donec aliud super hoc duxerimus ordinare : tibi committimus et mandamus, quatenus is indicta forma publices, seu publicari facias locis et temporibus opportunis, et nos inde certifices, cum super hoc fueris legitime requisitus. Dat apud Charth. 9 Cal. Julii, anno Dom. MCCC. consecrat nostrae VI.—*Wilkins' Concilia.*

at no great distance from the ground, and frequently immediately beneath a large window. These low side windows, or the lower portions of them, we commonly find closed up with masonry; and, on examination, they appear not to have been glazed, but externally covered with an iron grating, with a wooden shutter, opening inwardly, the hinges of which are frequently left im-

Low Side Window, Dallington Church, Northamptonshire.

bedded in the masonry, though the wooden shutters seldom remain. These I consider to have been confessional windows used by the friars, and those alluded to by Thomas Bedyll, Clerk of the Council in the reign of Henry VIII., and one of the Commissioners at the visitation made on the suppression of religious houses and chantries, who, in a letter to Crumwel, says,—"We think it best that the place wher thes frires have been wont to hire outtward confessions of al commers at certen tymes of the yere be *walled* up and that use to be for-doen for ever."ᵘ

In the south aisle of Kenilworth church, Warwickshire, there was, a few years ago, one of these windows, where the wooden shutter was remaining. This has been destroyed. On the south-west side of the chancel, which is decorated, of Lyddington church, Rutlandshire, is a low side window of one light, divided by a transom; of this the lower division is covered with an iron grating. Under the south-west window of the chancel of Cortlinstock church, Nottinghamshire, is a small square low side window, guarded by two upright and two transverse iron bars. On the south side of the chancel of Montacute church, Somersetshire, is a low side window. This has had a shutter on the inside, the iron staple of which still remains. It is impossible for a person outside the church to see the altar through this window. At the south-west of the chancel of Lyddington church, Rutlandshire, is a low side lancet-shaped window, trefoiled in the head with a hood mould over; this window is divided by a transom, the lower division being barred or grated. On the south side of the chancel of St. Margaret church, Dover,—a Norman structure,—and near the south-west angle, is what ap-

---

ᵘ *Letters Relating to the Suppression of Monasteries. Camden Society.* p. 47. These windows have been by some called Leper Windows, but I can find no authority for such a supposition.

pears to be an early example of the low side window. This is partly destroyed; but from the remains of the arch it appears to have been segmental, two feet six inches in width, one foot nine inches in height in the centre, and one foot three inches at the jambs, splayed internally, and the inner edge of the arch chamfered; the sill of the window is from three to four feet above the soil. In the third and westernmost bay externally of the chancel of Bilton church, Warwickshire, a chaste decorated structure of the fourteenth century, in the south wall beneath the westernmost principal light of a decorated window, with flowing tracery in the head, is a small square-headed window, now filled up with masonry; the hood mould of this is continued down vertically a short distance without a return. The internal arrangement of this window cannot be ascertained, as the wall is covered with modern wooden panel work.

Sometimes peculiar features may be observed in the arrangement of the interior of the low side window. In the north aisle of Doddington church, Kent, is a lancet-arched cinquefoiled-headed window; the upper part of this is barred with iron, but the lower part is blocked up with masonry, and hinges of a wooden shutter to this part are apparent; eastward of this, in the thickness of the wall, facing west, is a bracket and recess for an image, and beneath this a small stone desk; whilst opposite to this is a locker or ambrie, the door of which is gone, the staples remaining. These small blocked-up windows are by no means uncommon; they are mostly found in churches of the fourteenth century, sometimes earlier, but rarely in churches of the fifteenth century. One of the latest examples I have met with is on the south side of the chancel of Frowlesworth church, Leicestershire, a structure apparently late in the fifteenth century; this low side window, now blocked up, is square-headed, and of two lights, divided by a mullion. These windows are of extreme interest, and deserve minute examination, not only externally, but also, when possible, internally.

CHANTRIES, AND CHANTRY AND OTHER SUBORDINATE ALTARS. Besides the High Altar at the east end of the chancel, if a mere parochial church or chapel of ease, or, in the case of cathedral or conventual churches, at the east end of the choir, there were in most churches at least one or more altars, sometimes several, at which private mass could be celebrated, the officiating priest being the sole recipient of the mystical elements. In cathedral and large conventual churches these subordinate altars were very numerous, and dedicated to or called after the names of some saint or saints. In Durham Cathedral there were no less than twenty or thirty, or perhaps more, of these altars. In parochial churches we find traces of the existence of many of these altars, which, though generally destroyed, are indicated by the piscinæ which invariably accompanied them. These evince the former existence of altars in porches, perhaps rarely, as in the churches of Grantham, Notts; Melton Mowbray, Leicestershire; Pulham, Dorsetshire; and at the Cathedral of Durham. In crypts, as in the chapel of the Pyx, Westminster Abbey, and at Grantham, the ancient altars in which are still existing;

whilst at Repton and elsewhere the former existence of such is apparent. In vestries, as at Durham; Adderbury, Oxfordshire; and Warmington, Warwickshire; the two latter being still retained. At the east ends of aisles, before rood lofts, and apparently on them. In upper chapels, as in Gloucester Cathedral. In anchorages, as at Durham Cathedral. And in Capellæ carnariæ, or charnel vaults; and at the west end of tombs. But the most numerous of these altars were those pertaining to chantries, whereat private masses for the dead were celebrated.

To the prayers for the faithful and departed,—an early practice of the Church, as the Clementine and other most ancient liturgies evince,[r]—but which prayers contain no allusion to temporal pains in a future state, were at a later period, not for six hundred years after Christ, saith Bishop Jewell, superadded private masses for the dead, and prayers for deliverance of their souls from such pains as were founded on the doctrine of purgatory.

Of the old Fathers of the Church, Origen, circa A.D. 240, taught, in accordance with the views of Plato,

---

[r] Let us pray for all those who have fallen asleep in the faith.—*Clementine Liturgy*. Translation.

Give rest to the souls of our fathers and brethren that have heretofore slept in the faith of Christ. O Lord our God.—*Liturgy of S. Mark*. Translation.

For repose and remission of the soul of thy servant, *N.*, in a place of light, where sorrow and sighing are put away. Give him rest, O Lord our God.—*Liturgy of Basil*. Translation.

And to the spirits of all these give rest.—*Liturgy of S. Mark*. Translation.

And remember all those that are departed in the hope of the resurrection to eternal life, and give them rest where the light of Thy countenance shines upon them.—*Liturgy of S. Chrysostom*. Translation.

that the souls of good men would hereafter, namely, at the Day of Judgment, pass through a purgatorial fire.

St. Augustine, circa A.D. 407, thought it probable that the purgation of souls by fire, which Origen had taught, might take place in the interval between death and the Day of Judgment, instead of being deferred until the latter period. Thus advancing on the views of Origen.

Gregory the Great, circa A.D. 590—604, in his Dialogues, being asked the question, *Doceri vellem si post mortem purgatorius ignis esse credendus est?* replies, *De quibusdam levibus culpis esse ante judicium purgatorius ignis credendus est.*⁹ He is reputed by some to have first established this doctrine as an Article of Faith; and it is probable that it was first brought through his missionaries into Britain, for we have no evidence that the early British Church was acquainted with such a doctrine.

Venerable Bede, who flourished A.D. 716—735, in a passage in which he describes, according to the then belief, the state of souls in purgatory, which would be redeemed at the Day of Judgment, adds, "*Multos autem preces viventium et eleemosynem, et ieiunia, et maxime celebratio missarum, ut etiam ante diem judicii liberentur adjuvant.*"ᶻ

Matthew Paris, the historian, who flourished in the thirteenth century, writing of this doctrine, observes,— "*Per missas vero, psalmos, eleemosynas, et orationes ecclesiæ generalis, et per specialia amicorum auxilia, aut*

---

⁹ *Dialogorum, lib. IV. c. xxxix.*
ᶻ *Eccl. Hist. lib. V. c. xiii.*

*Purgandorum tormenta mitigantur, aut de ipsis suppliciis ad minora transferuntur donec pœnitus liberentur.*"ᵃ

This doctrine was still more explicitly defined by the Council of Florence, held A.D. 1438, which upheld it as an Article of Faith, declaring, "That the souls of true penitents dying in the love of God, before they have brought forth fruits worthy of the repentance of their sins, are purified after their death by the pains of purgatory; and that they are delivered from these pains by the suffrages of the faithful that are living, such as holy sacrifices, prayers, alms, and other works of piety which the faithful do for the other faithful, according to the orders of the Church."ᵇ A doctrine, however, rejected by the Anglican Church as "a fond thing vainly invented, and grounded upon no warranty of Scripture, but rather repugnant to the Word of God."

Throughout the Middle Ages, however, we find this doctrine prevalent: of its effect in Anglo-Saxon times in the internal arrangement of our churches, we have hardly existing remains enough to judge; but in the great Benedictine churches erected in the twelfth century, we meet with numerous apsidal chapels surrounding or annexed to an apsidal choir, each of which contained an altar; and in the conventual churches of the Cistercian Order we find the transepts on the eastern side contained altars, each transept being divided into two or three compartments, with an altar in each for the celebration of private mass, in which masses that of *Missa pro defunctis* was of all others, perhaps, the most accustomed.

[a] *Hist. Angl. Edit. Watts*, p. 76.
[b] *Dupin's Eccl. Hist.*, Vol. XIII., English Translation, p. 45.

The most ancient of our Monastic Establishments, those of the Benedictine Order, were founded between the seventh and twelfth centuries, but the churches of these were mostly rebuilt sometime during the twelfth century. The Cistercian Order, a reformed emanation from the Benedictine rule, was introduced into this country about the fourth decade of 'the twelfth century, circa A.D. 1130—1135, when numerous monasteries of that Order were founded by rich and powerful individuals and plentifully endowed, not only by the founders, but by others who with grants of lands bequeathed their bodies for burial in some particular monastic church, in order that remembrances might be made of their souls and masses daily celebrated therein for the deliverance of the testator's soul from purgatory.

Besides the two great Monastic Orders, the Benedictine and the Cistercian, other Orders of a minor description were founded, of which the most noticeable were those of the Friars of the Orders of St. Francis and St. Dominic. In the thirteenth century the foundation of religious communities on a large scale mostly ceased, and the foundations of chantries on a lesser scale commenced. These were endowed for the maintenance of a priest to celebrate, daily, mass at some particular altar in a church for the benefit of the souls of the founder, his family, and friends.

In many, if not most, of our parish churches with aisles we commonly find indications of the former existence of an altar at the east end of each aisle, by the piscina with its fenestella or niche, ofttimes canopied, in the south wall near the east end, and which fenestella sometimes contained a shelf of wood or stone for the reception of the Eucharistic vessels. This portion of the aisle was often separated from the rest of the church by a screen, sometimes of stone but generally of wood, the lower part of close panel-work, the upper part of open-work tracery, similar to that forming the division between the chancel and nave; and the space thus enclosed was converted into or became a private chapel or chantry, endowed with land sufficient for the maintenance of a priest or more than one, for which endowment licenses from the Crown had to be obtained, the statutes of mortmain forbidding such without. The chantry priest was required to celebrate private masses daily or otherwise as the endowment expressed, at the altar therein erected and generally dedicated to some Saint, for the souls of the founder, his ancestors, family, and posterity, for whose remains these chantry chapels were frequently appropriated as burial places.[c] At this service, however, no congregation was required to be present, but merely the priest and an acolyte to assist him, and it was in allusion to the low or private masses thus celebrated, that Bishop Jewell, whilst condemning the practice as untenable, observes, "And even suche be their private masses for the most part, sayde in side iles, alone, without companye of people, onely with one boye to make answer." And Fox, the martyrologist, commenting on the same practice, sarcastically observes,—"Sir John (*i.e.* the priest) is kin to the tide, he will tarry for no man, if he have a boy to say

[c] The earliest instance I have met with of the foundation of a chantry is one founded by King John, A.D. 1204.—*Monasticon, Vol. 2, p. 337.*

amen it is enough." When the general though not universal destruction of the chantry altars took place, the piscina, indicative of the site of the chantry chapel, was suffered to remain, and oftentimes the ambrie or locker, though we do not find the latter so frequently as the former. Sometimes, when the foundation was considerable, we find the sedilia or stone seats for the three clerical orders, *viz.*, the priest, deacon, and subdeacon, who officiated at high mass, in the south wall of the aisle or chantry chapel, as in the south aisle of the church of Stratford-upon-Avon, Warwickshire, and in the south wall of a chapel on the south side of the chancel of Great Yarmouth church, Norfolk. The screens by which these chantry chapels were enclosed have in numerous instances, especially of late years, been destroyed; still many have been preserved, and chantry chapels parted off the church by screen-work of stone may be found in the churches of Bradford Abbas, Dorsetshire, and Aldbury, Hertfordshire; in which latter church is a very perfect specimen of a mortuary chapel, containing in the midst of it a high tomb with recumbent effigies. Chantry chapels enclosed in two of the sides by wooden screen-work are more common.

Besides the chantry chapels at the east end of an aisle parted off by screen-work, it was by no means unusual in the fourteenth and fifteenth centuries, for individuals to erect distinct chantry chapels, annexed to a church, that is separated from the church by a single or two arches, with panel screen-work carried across the arches. In these several chapels the founder would sometimes erect his tomb with his effigy thereon. Sometimes an altar was attached to a tomb. To the west end of one of the tombs in Arundel church, Sussex, a small altar forms an adjunct. John, Lord Marney, by his will dated 10th March, 1524, appoints his own burial in the middle of the new isle or chapel attached to the church of Layer Marney, in the county of Essex, the tomb to be of such stone as his father's or else of grey marble. He appointed an image of brass for himself, and on either side one for each of his two wives; also that at the west end there should be an altar for a priest to sing for him perpetually. A mass of masonry, the remains of this altar, from which the covering slab or altar stone has been removed, still exists as an adjunct at the west end of this tomb.

There is a peculiar class of chantry or sepulchral chapels which formerly contained small altars, namely, those constructed in several of our cathedral and conventual churches, chiefly erected by ecclesiastics of episcopal rank, in which their bodies were deposited, sometimes by members of the laity, and which contained their individual tombs and effigies thereon. These chantries were stone structures designed and executed with much taste and skill and placed between two piers, the lower portions being of blank panel-work, the upper portions of open screen-work, whilst the roof was more or less elaborately groined. These present specimens of the most tasteful and enriched designs it is possible to conceive, and much of the finest sculptured detail was combined with the architectural features there displayed. Of these small insulated and mortuary chapels may be noticed those in Winchester Cathedral, of Bishop Edington, who died A.D. 1366; of Bishop William of Wyckham, who died A.D. 1404; of Cardinal Beaufort, who died A.D. 1447; and of Bishop Wainflete, who died A.D. 1486. Wells Cathedral contains the insulated chantry chapels of Bishop Bubwith, who died A.D. 1424, and of Dean Sugar, who died A.D. 1489. Tewkesbury Abbey church contains the insulated chantry chapel of Isabel, Countess of Warwick, erected A.D. 1438. Worcester Cathedral contains that of Prince Arthur, who died A.D. 1502. St. George's chapel, Windsor, the Oxenbridge chantry, A.D. 1522. Salisbury Cathedral, that of Bishop Audley, who died A.D. 1524. Gloucester Cathedral, that of Abbot Parker, erected between A.D. 1515—1539: whilst the Priory church of Christchurch, Hants, contains the insulated chantry chapel erected early in the sixteenth century by Margaret, Countess of Salisbury.[d]

Of altars still existing in their original positions the following may be noticed. In the chapels near the whispering gallery, Gloucester Cathedral, and up aloft, one altar is complete, and another nearly so, on the slab of one three of the five incised crosses are still visible. In a crypt on the north side of the chancel of Solihull church, Warwickshire, is an ancient stone altar, consisting of a large slab of considerable thickness supported by three plain moulded brackets projecting from the wall, on which three of the five crosses with which the

---

[d] An illustrated work, with plans, elevations, and details, of these several interesting and insulated chantry chapels, is a desideratum, and would form a most valuable addition to the various works we possess on medieval architecture.

slab was marked, symbolical of the five wounds of our Saviour are still visible, and to the south of this altar is a piscina. Beneath the chancel of Bedale church, Yorkshire, is a crypt, a structure of the fourteenth century, in which the original stone altar, marked with its five crosses, is still existing.

The high altar is still remaining at the east end of the apse, Peterchurch, Herefordshire, with the five crosses incised upon the slab; and on either side of the entrance into the apse is a stone altar, probably the rood loft altars.

Rood loft altars, or as they were sometimes designated, altars of the crucifix, were anciently very numerous; they were placed either westward of, and adjoining the rood loft screen, or up aloft, and probably on the rood loft itself. Besides the altars at Peterchurch, the only rood loft altars I have met with yet existing in this country are two beneath the rood loft in the little church of St. Patricio, near Crickhowel, South Wales; one placed on each side of the entrance into the chancel, westward of and against the screen supporting the rood loft. Both of these altars are of plain masonry, with the usual thick projecting covering slabs or altar stones, each marked with the five crosses, and the under part of each slab chamfered. On either side of the entrance into the chancel, Urishay church, Herefordshire, is a stone altar.

We sometimes meet with allusions to these altars. Gervase, in his account of the destruction and reparation of Canterbury Cathedral in the latter part of the twelfth century, has a passage which, I think, alludes to one of these, the altar of the cross, when he says, "*Pulpitum vero turrem predictam* (he is speaking of the central tower) *a navi quodammode separabat, et ex parte navis in medio sui altare sanctæ crucis habebat.* Browne Willis, in his *Survey of the Cathedral Church of S. David,* South Wales, writing of Richard de Carew, Bishop of that See from A.D. 1256 to A.D. 1280, tells us he was buried in his own cathedral, near the altar of the holy rood, on the south part. This must have been under the rood loft, near the door leading from the nave into the quire; and Leland, in his *Collectanea,* informs us that this bishop was buried near the altar of the crucifix,— *Nomina episcoporum sepultorum in Meneven: eccl:— Ricardus Carew prope altari crucifixi.*

In the account of the Lancastrian chantries, published by the Chetham Society, we find a few references to these altars.

Parish church of Croston, "The Chantrie at the roode altar within the aforesaid church."

Parish church of Standycke, "The Chauntire at the rode alter within the p'oche church biforsayde."

"The Chauntire at the alter of the crucifixe w<sup>t</sup>hin the p'oche church of Preston."

Of altars formerly existing beneath or over rood lofts, I may now notice the *indicia*; at Burg church, Herefordshire, in the south wall, high above the rood loft, is a piscina, indicative of an altar appended to the rood loft. At Wigmore church, Herefordshire, on the south side of the nave, high up in the wall, is a piscina. At Maxey church, Northamptonshire, on the south wall of the clerestory is a piscina. At Deddington church, Oxfordshire, in the east wall of the nave, south of the chancel is a piscina. At Eastbourne church, Essex, in the south wall of the nave is a piscina. In Tenby church, South Wales, in the south wall of the south aisle close to where the rood-loft was carried across, is a piscina. In the south wall at the east end of the nave of Bolton church, Yorkshire, and beneath where the rood-loft had been apparently placed, as there were projecting brackets for the support of a soller, is a piscina indicative of a rood-loft altar. The piscinæ of rood-loft altars have been lately brought to light in the reparations of the chapel of Brownsover, and in the churches of Bilton, Chesterton, and Church Lawford, Warwickshire. Many other like examples of the former existence of rood-loft altars will doubtlessly be brought to light.

Both the high altars, and also the chantry and other altars, were generally constructed of a solid mass of masonry covered with a slab or table of stone, in most instances six inches in thickness, the lower three inches being chamfered to that extent. An incised cross appears in the centre and near each angle of the table or slab. Altars thus constructed, besides those previously noticed, are existing in the north aisle of Bergeworth church, near Evesham; and in the south aisle of Enstone church, Oxfordshire.

In the Maison Dieu, at Ripon, Yorkshire, is a stone altar, the covering slab of which is 5 feet 9 inches long, 2 feet 7 inches wide, and 6 inches thick, with the under edge plainly chamfered. In the south aisle of the choir, St. Alban's Cathedral, is what appears to be a plain high tomb, but apparently an ancient altar, transposed for preservation; the covering slab is of Durham marble, 5 feet 10½ inches long, 2 feet 9¾ inches wide, and 8 inches in thickness; the under part being hollowed; on the surface the five incised crosses are visible.

In a small chantry chapel, or vestiarium, on the north side of the chancel of Claypole church, near Newark, an ancient stone altar, which for preservation had been covered with planks of wood, was some years ago brought to light. The altar slab is supported on three distinct masses of masonry with recesses between, and four of the five incised crosses are remaining on the surface. The length of the slab is 7 feet 2 inches—the width of the chapel; the breadth of the slab, 3 feet 2 inches, and the thickness 4½ inches. At Tarring Neville church, Sussex, is a perfect chantry altar at the east end of the south aisle; the five crosses are, however, all but obliterated. In Arundel church, Sussex, are three stone altars.

Although plain masonry formed the rule in the construction of altars, there were exceptions. In the oratory or private chapel in the prior's lodgings, Wenlock Abbey, the altar still remains; it is panelled in front. Sometimes the altar consisted of a thick stone slab or table, with an incised cross on the surface at each angle, and in the centre, supported merely on brackets or trusses of stone built into and projecting from the wall, as in the revestry, Warmington church, Warwickshire; close to this, in the east wall, is a piscina.

In the ancient vestiarium on the north side of the chancel of Shoteswell church, Warwickshire, is an ancient altar, like that at Warmington in the same county, projecting from the east wall and supported by brackets. At Belper chapel, Derbyshire, is a stone altar in its original position, fixed in the wall of the chancel immediately beneath the east windows. In the church of Llandegfan, near Beaumaris, Anglesea, is

Altar, Warmington Church.

a stone altar slab resting on brackets beneath the east window. In the vestry of Adderbury church, Oxfordshire, is an ancient stone altar constructed of solid masonry, the covering slab is 7 feet 6 inches long, and $2\frac{1}{4}$ inches thick; this is in its original position, beneath a bay window looking eastward. In the chapel of the

Vicar's close, Wells, the altar is still existing, and retains its original slab marked with the five crosses.

In a small chapel in the roof of Compton Wyniate House, Warwickshire, erected in the reign of Henry VII., the window-sill of *wood*, marked with the five crosses, indicates it to have been used as an altar. Near to this chapel is a place of concealment for a priest.

Ancient altar stones or tables, removed from their original position, are to be found on the pavement of many of our churches. A few instances may be enumerated. In Howell church, Lincolnshire, is an altar slab bearing the five crosses: close by, and projecting from the east wall, are three brackets which supported it. At Ranceby church, Lincolnshire, an altar slab lies within the altar rails. At St. Clement's church, Sandwich, Kent, is an altar slab forming part of the pavement; this is 8 feet 3 inches in length, and 4 feet in width; the five crosses on it are visible. In Whitwell church, Northamptonshire, are two altar slabs, one of which is broken, in the chancel, the other on the pavement near the chancel arch; on each of these the five crosses are visible. In a chapel at the east end of the north aisle of St. Mary Magdalen church, Wigenhall, Norfolk, is an altar slab with the crosses remaining. On the chancel floor in Freshlingfield church, Suffolk, is an altar stone. In South Collingham church, Notts, in the pavement of the chancel appear fragments of the ancient altar slab. In the churches of Maidstone and of Lenham, Kent, the altar slabs form part of the pavement of the chancels; on each are remains of the incised crosses. On the pavement of the south aisle of Brixton church, Isle of Wight, is an ancient altar slab of purbeck marble bearing the five crosses: it is 6 feet 2¼ inches long and 1 foot 9½ inches wide. Lying in the north aisle of All Saints' church, York, is an altar slab. In the church of St. Michael le Belfry, York, one of the altar slabs forms part of the pavement; it was used for a sepulchral slab for one of the Mayors in 1570, and again for another person in 1746.

The above enumeration of despoiled altar slabs is by no means exhausted, and might easily be added to, as the examples are very numerous. It might well have been supposed that as many of the altars were taken down unwillingly, the altar slabs bearing the incised crosses would have been placed on the pavement with the surface downwards; but as the lower edges were bevelled or chamfered this could not be done, since the pavement would have been rendered irregular—hence the position in the pavement with the crosses uppermost.

RELICS INCLOSED IN ALTARS. Although many if not most altars had relics deposited in them, an altar might be consecrated without relics, and in Bishop Lacy's Pontifical the form " Consecratio altaris " makes no mention of the deposition therein of relics.[e] There is, however, another form in the Pontifical " Reconditio reliquiarum " in which the confessio or sepulchrum is set forth as the place where the relics were deposited, and in one part of the service is the following rubrical direction: Si reliquie non habentur omittendum est officium illorum.

[e] *Liber Pontificalis* of Edmund Lacy, Bishop of Exeter, A.D. 1419—1455. The Pontifical, however, is supposed to have been written in the fourteenth century.

Lyndwood, in his gloss on one of the Provincial Constitutions of Stephen, Archbishop of Canterbury, A.D. 1207—1228, "De reliquiis et veneratione sanctorum," observing on the words "Vetera quoque corporalia que fuerint non idonea in altaribus quando consecrantur loco reliquiarum reponantur, vel in presentia Archidiaconi comburantur," tells us that altars ought not to be consecrated without relics, but if an altar should be consecrated without the consecration would not be void. Hence it appears that relics were not essential to the consecration of an altar.*f* Sometimes the relics were contained in a small coffer of stone built up into the altar. In the reparation of Ashbourne church, a few years ago, in removing the soil at the east end of the chancel where the high altar formerly stood, a small coffer of stone of an oblong shape about a foot in length was discovered. This I imagine to have been the sepulchre or confessio within which the relics pertaining to the structure of the altar were inclosed and concealed.

HAGIOSCOPE. In several of our churches we find an opening or aperture obliquely disposed carried through the thickness of the wall at the north-east angle of the south aisle, and the south-east angle of the north aisle, or of chapels eastward of the aisles.; and which oblique apertures opened into the chancel. This opening was the *Hagioscope*,*g* through which at high mass the eleva-

*f* Loco reliquiarum sine quibus alteria consecrari non debent . . . . . Si tamen consecratur altare sine reliquiis tenet consecratio secundum Hugonem . . . . . . Unde licet reliquie non sint de substantia consecrationis altaris.—*Lyndwood, Provinciale, &c.*

*g* This is a modern term by which these oblique openings are generally known.

tion of the Host at the high altar and other ceremonies might be viewed from the chantry or other chapel, situate at the east end of each aisle, or contiguous to the chancel. In general, these apertures are mere plain, narrow, oblong slits; sometimes, however, they partake of a more ornamental character, as in a chantry chapel on the south side of Irthlingborough church, Northamptonshire, where the head of an aperture of this kind is arched, cinquefoiled within, and finished above with an embattled moulding. In the north and south transepts of Minster Lovell church, Oxfordshire, are oblique openings, arched-headed and foliated; and in the north of Chipping Norton church, in the same county, is a singular hagioscope, obliquely disposed, not unlike a square-headed window of three foliated arched lights, with a quatrefoil beneath each light. An aperture of this description is to be met with in a north chapel on the north side of the chancel of Standground church, near Peterborough. In a south chapel, eastward of the south aisle and adjoining the chancel of Marston church, Bedfordshire, is an aperture of this description. At St. Nicholas church, Marston, Oxfordshire, an oblique aperture is carried through the wall at the north-east angle of the south aisle, communicating with the chancel. At Beckley church, Oxfordshire, are two oblique apertures of this kind: one at the northeast angle of the south aisle, the other at the southeast angle of the north aisle. These apertures or oblique openings, though not general, are by no means uncommon in our churches.

ENSHRINED HEARTS, RELIQUARIES, ETC. Besides the

rich tablernacle-like structures of stone for the reception of shrines, containing the relics of saints, formerly existing in our cathedral and conventual churches, and of which some notice will be given, the walls of some of our churches have been found to contain enshrined hearts, and relics discovered on the removal of walling. These had sometimes rude sculptured designs in front; and some of these are still remaining intact. Some years ago a sculptured stone, representing two hands

Sculptured Stone, Yaxley Church, Huntingdonshire.

holding up a heart, was removed from the north wall—of the fourteenth century—of the south transept of Yaxley church, Huntingdonshire, and behind this a small cylindrical-shaped box, 4½ inches high, and 4 inches in diameter, with a turned cover, was found. This had doubtless contained a heart, which had decayed.

In Leybourne church, Kent, two depositories of stone are said to have been found in a niche in the north aisle; one of these contained a heart in a leaden box; the other was empty. The niche is of transition character from Early English to Decorated, and is now built into a wall of Perpendicular date; the depositories were of the same date as the niche.

In the south wall of the chancel of Burford church, Shropshire, near Tenbury, are two cylindrical hollows, each covered with a flat circular dish of stone. One of these formerly contained the heart of Edmund Cornwall, Knight, who died at Cologne, in the fourteenth year of the reign of King Henry VI.

Against the north wall of the nave of Adwell church, Oxfordshire, is a bust of a man represented as clad in chain mail; before it is a large shield, devoid of any heraldic device, but the two hands of the figure support, above the shield, what is apparently intended for a heart. At Hampton-in-Arden church, Warwickshire, was, until lately, in the south wall of the chancel, near the south door, a pointed arched recess, trefoiled within, and supported by shafts; on this was sculptured a demi-figure holding a shield, charged with two lions. This wall was taken down in the course of some recent reparations, but nothing was found behind this sculptured tablet. In the north wall of the chancel of Cuberly church, Gloucestershire, beneath a plain pedimental canopy, is a pointed trefoil-headed arch, supported on engaged shafts; within a vesica is the bust of a man in a coif and hawberk of mail of rings set edgewise, and in front of the breast is a shield, crossed by the arms of a figure, the hands of which appear as if holding up a heart. This tablet is a work of the fourteenth century, and is 3 feet 4 inches in height, by 1 foot 9 inches in width. Behind it the heart of a Lord Berkeley is said to have been deposited. In a wall of Woodford church, Northamptonshire, a heart was a few years ago discovered.

The heart of William Giffard, Bishop of Winchester, and founder of the Cistercian Abbey of Waverly, who died A.D. 1128—9, was found in taking down a wall at the north-west end of the church of that abbey, in a stone depository containing two leaden vessels, soldered together; the heart had been preserved with spices, and was not decayed. It was no uncommon practice for the hearts of persons of note to be consigned for sepulture to some church apart from that in which their bodies were interred. We also find concealed in the walls of churches small reliquaries, containing remains esteemed to be sacred. St. Gregory the Great, who flourished at the close of the sixth century, is said to have regarded relics with great veneration, and to have made use of them in the consecration of churches; he also sent divers relics into Britain to his missionaries, Augustine and Mellitus.

A small stone reliquary, containing a bone, was some years ago found in the south wall of the south aisle of Brixworth church, Northamptonshire. It still remains in that church, and is about 15 inches high by $7\frac{1}{2}$ inches wide; at the angles are shafts with bases and capitals. Three of the four faces present trefoil-headed panels, with pedimental canopies enriched with crockets; at each angle a pinnacle was apparently fixed, the mortice holes of which are still apparent. On removing the top of this reliquary—for it is composed of two parts—a cavity was disclosed about 4 inches in diameter, and about the same in depth; the back of the reliquary also presents this cavity to view. This reliquary is of the fourteenth century.

 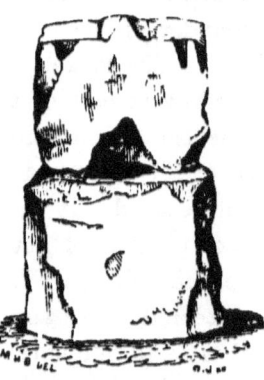

Reliquary, 14th Century, Brixworth Church, Northamptonshire.

About the year 1849, in taking down the north wall of the nave of Kew Stoke church, near Weston-super-Mare, Somersetshire, it became necessary to remove a block of stone sculptured with a demi-figure, placed in a niche, which was built into the wall below the sill of a window on the inside of the church. In the back of this block was hollowed out a small arched recess, within which was deposited an oaken vessel or cup, partially decayed, and a little split open. In the bottom of this cup was a dry black incrustation of what appeared to

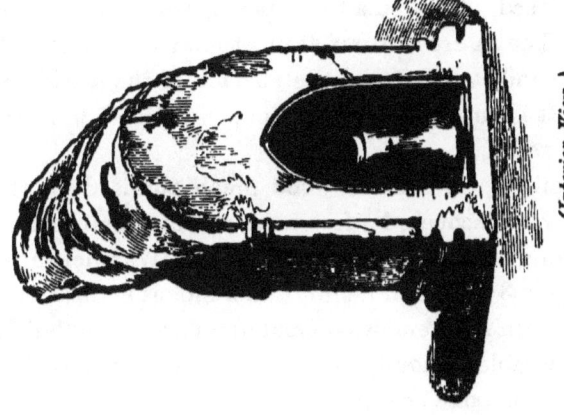
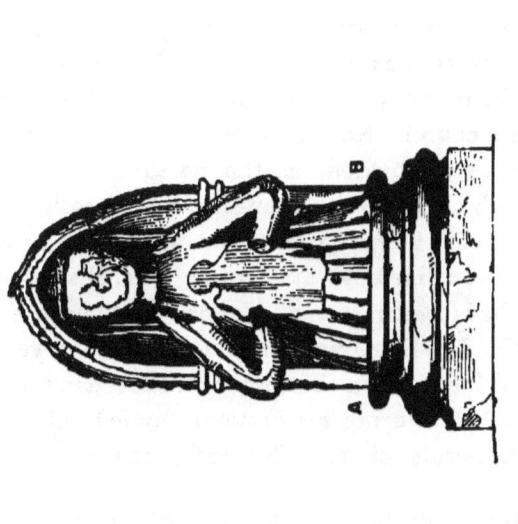

Reliquary, Kew Stoke Church.

(Exterior View.) (Interior View.)

have been coagulated blood. The cup had a rim at the top, as if to receive a cover; the cavity in the stone was firmly closed with a small oak panel, which fitted to a rebate. The figure in front seems to have held a shield, but the hands are lost and the surface of the shield defaced. It is supposed that this cup was the depository of some relic.

A small cup of bone or wood much abraised on the surface was discovered a few years ago on some reparations being made in Lapworth church, Warwickshire. This cup is 3 inches in height and 2 inches in diameter: at the mouth it resembles a small drinking horn, but is of considerable antiquity, and had possibly been used to contain some bone or relic.

EARTHEN JARS. Church walls have been found sometimes to contain empty earthenware jars. It is supposed they were thus placed for acoustic purposes.

In or about the year 1864, in repairing the chancel of Denford church, Northamptonshire, some singular orifices were discovered on the north side over the arcading which lines the chancel, and beneath a discharging arch. These orifices had all originally contained or been lined by earthenware pots, the bottoms of which had been removed. In one of them the pot still remained.[h] Some years ago some earthen vessels were found immured in the walls of a church in Oxfordshire.[i] In 1852, during restorations carried out in St. Peter's Mancroft church, Norwich, the remains of

[h] *Reports and Papers of Architectural Societies*, vol. 7, p. 102.
[i] *Journal of the Royal Archæological Institute*, vol. 7, p. 314. The name of the church is not given.

passages were found under the chancel floor, having earthen jars imbedded in the side walls. The vessels of red ware with a slight glaze on the upper part, were laid horizontally, about 4 feet apart, their mouths being flush with the face of the wall; they measured 8 inches in height, and the diameter of the mouth was about 6 inches. At Fountains Abbey vessels imbedded in masonry below the level of the floor at the east end of the nave have been discovered. Vestiges of a similar passage under the chancel floor, in the side walls of which several one-handled jars or pitchers were found imbedded, came to light some years ago in the reparation of St. Nicholas' church, Ipswich.[k] In 1747, on rebuilding Fairwell church, Staffordshire, there were found in the south wall about 6 feet from the ground, three ranges of coarse earthen vessels of different sizes and unglazed, the largest capable of containing two, the smaller one quart. The larger vessels were $4\frac{1}{2}$ inches across the mouth, 24 inches in circumference, and about 1 foot in height. The smaller were 3 inches across the mouth, 16 inches in circumference, and about 6 inches in height. They lay on the sides in one direction, their mouths placed towards the inner side of the church, and stopped with a thin coat of plaster.[l] Empty vessels of pottery have also been found, immured in the walls of other churches.

DEDICATION OR CONSECRATION CROSSES. We sometimes meet with dedication or consecration crosses imbedded in the external walls of churches. Three

[k] *Journal of Archæological Institute, vol.* 12, *p.* 276—7.
[l] *Gough's Sep. Mon. vol.* 1, *p.* 16.

of these crosses appear on one of the walls of Strixton church, Northamptonshire; and, let into the east jamb of the south doorway of Elton church, in the same county, is a small diamond-shaped piece of purbeck marble, of about the size of a window quarry; on this a cross fleury has been sunk. In one of the walls of Cockerington church, Lincolnshire, is a dedication cross. At Uffington church, Berkshire, the consecration crosses are sunk in stone quatrefoils. Worked up into the wall of the north aisle of Chedzoy church, Somersetshire, are crosses within circles, terminating in rosettes, with a rosette in the centre. A like cross also appears on the wall of the south aisle of the same church. Other churches will be found to display one or more of these consecration crosses imbedded in the external walls.

THE VESTIARIUM OR VESTRY. This, as thus defined by Durandus,—"*Sacrarium sive locus in quo sacra reponunt, sive in quo sacerdos sacras vestes induit,*"—was a small apartment generally placed on the north side of the chancel, and communicating therewith by an internal doorway, but had no external door. In many instances the vestry was a subsequent adjunct to the chancel. Very rarely the ancient vestry occurs on the south side of the chancel, as at Farcett church, Huntingdonshire; but this was an unusual arrangement, and it was entered through a door in the south-east wall of the chancel. It is 12 feet in length from north to south, and 5 feet 7 inches in width. The vestry—of the fifteenth century—of Brightlingsea church, Essex, is also on the south side of the chancel. At Paignton church, Devon, the vestry is also on the south side of the chancel, but instances of its being thus placed are comparatively few. Sometimes the position of the vestry or revestry was at the east end of the chancel, behind the high altar. This is the case at Kingsbury episcopi, Somersetshire, where the revestry behind the altar is entered through a door in the east wall at the south end. The same position of the vestry occurs at Langport church, in the same county, where it is entered through a square-headed doorway; and at the Beauchamp chapel, St. Mary's church, Warwick. At Winchcomb church, Gloucestershire, the design was to have had a vestry behind the altar, the stone ties being left in the east wall, but this intention appears to have been never carried out. In the vestry were kept the vestments and other articles; and in it we find, occasionally, an altar, with its piscina, and a chamber over.

THE CHURCH CHEST. This, an article of church furniture rarely indeed enumerated or enjoined as necessary by any episcopal injunctions, but yet of considerable antiquity, we find generally placed in the vestry, but sometimes in the tower or some other part of the church. It appears anciently as an offertory coffer. By a mandate issued by King Henry II., A.D. 1166, for contributions towards the defence and assistance of the Christians in the Holy Land, a coffer, *truncus*, was enjoined to be placed in every church; which coffer was to have three keys, of which the priest was to keep one, whilst the most trustworthy of the parishioners were to have the custody of the other two. "*Et erit truncus in ecclesia episcopali, et per singulas villas in ecclesiis* . . . . . . .

*Truncus vero habebit tres claves, quarum unam custodi et Presbyter duas fideliores viri de parochia.*"

Another instance of a general mandate for setting up these coffers in churches, though, like the former, for a special purpose, is noticed in a letter of Pope Innocent III., who, A.D. 1200, when about to tax the Church under the ostensible object of providing means for the benefit of the Holy Land, wrote to the Archbishop and Bishops of the different provinces and dioceses, in which letter occurs the following passage:—" To this end we command that in every church there shall be placed a hollow trunk, fastened with three keys, the first to be kept by the bishop, the second by the priest of the church, and the third by some religious layman; and that the faithful shall be exhorted to deposit in it, according as God shall move their hearts, their alms for the remission of their sins. And that once in the week in all churches mass shall be publicly sung for the remission of sins, and especially of those who shall thus contribute." "*Ad hæc in singulis ecclesiis truncum concavum poni præcipimus tribus clavibus consignatum, prima penes Episcopum, secunda penes ecclesiæ sacerdotem, tertia penes aliquem religiosum laicum conservandæ, et in eo fideles quilibet, juxta quod eorum mentibus Dominus inspiraverit, suas eleemosynas deponere in remissionem suorum peccaminum moveantur; et in omnibus ecclesiis semel in hebdomada pro remissione peccatorum et prosertim offerentium missa publice decantetur.*""

In the curious and interesting chronicle of Jocelin

---

" *Chronica Gervasii. An. Græ.* 1166.
" *Rogeri de Hoveden Annalium, sub anno* 1200.

of Brakelond, a monk of St. Edmundsbury, detailing numerous matters which occurred at that monastery, and of which the writer was personally cognizant, from A.D. 1173 to A.D. 1202, we have another purpose explained to which offertory boxes were anciently put. He tells us that Warin, a monk, the keeper of the shrine of St. Edmund, and Sampson, the subsacrist, made a certain hollow trunk, with a hole in the middle or at the top, and fastened with an iron lock: this they caused to be set up in the great church, near the door without the choir in the way of the people, so that therein persons should put their contributions for the building of the tower. "*Truncum quendam fecerunt concavum et perforatum in medio vel in summo, et obseratum sera ferrea; et erigi fecerunt in magna ecclesia, juxta hostium extra chorum in communi transitu vulgi, ut ubi ponerent homines elemosinam suam ad edificationem turris.*"°

There often appears to have been offertory boxes affixed at or near to the shrines or images of those who had been canonized, or were held in veneration as Saints or Martyrs. The offertory box called the Pix of St. Cuthbert, at the shrine of that saint in Durham Cathedral, was secured by two locks. From accounts preserved of the receipts from this box for 9 years, between 1378 and 1388, the average for each year appears to have been £24. 10s. 0d., a sum equivalent at the present day to an income of nearly £500. per annum.

An offertory box, which was probably placed near the image or tomb of St. Godric, is mentioned or alluded to in the compotus or account Roll of Finchale

° *Chronica Jocelini de Brakelonda, p. 7* : edited for the Camden Society.

Priory, near Durham, sub anno 1355, and it is there called the "Pixis." "*Et de Lvs. v.d. ob receptis de picside Sancti Godrici per diversas vixes.*" And under the year 1363—4, "*Item respondet de xiiis. xd. receptis de Pixide Sancti Godrici.*"

Of the stone offertory box in Bridlington Priory church, Yorkshire, I have before treated, at page 63.

There is one other stone offertory box I know of, still existing. This is the hollow stone bracket affixed to the monument of Edward II. in Gloucester Cathedral. Miracles are said to have occurred at the tomb of this king, and the oblations thereat were so great, that the choir is said to have been vaulted from the proceeds of the offerings of the faithful flocking to the tomb. This work is recorded to have been accomplished during the Abbacy of Adam de Staunton, between A.D. 1337 and A.D. 1351. "*Cujus tempore constructa est magna volta chori magnis et multis expensis et sumptuosis cum stallis ibidem in parte prioris ex oblatione fidelium et tumbam regis confluentium.*"

But in general the church chest, as required by the Synod of Exeter, A.D. 1287, was for the custody of the books and vestments. Sometimes we find it rudely formed or hollowed out of the solid trunk of a tree, with a plain or barrel-shaped lid of considerable thickness. These primitive-looking chests are often strongly banded about with iron. In Whitwell church, Rutlandshire, is a chest which has been hollowed out of solid timber, and is strongly bound with iron. In St. Martin's church, Leicester, is a chest of the same description, but the dimensions are larger. In St. Margaret's church,

Leicester, is a chest formed or hollowed out of a solid log; the lid is covered with iron-work, in which are two apertures for the reception of money, and the chest has three locks and hinges. At Churchill, Worcestershire, the parish coffer, kept in the church, is of singularly rude workmanship, consisting merely of a huge block of oak, hollowed, with a thick cover. In the vestry of Bampton church, Oxfordshire, is a curious trunk chest. In the south aisle of Wimborne Minster is an old oak chest, or rather an oak tree excavated, to which six massive locks were formerly affixed. In Ensham church, Oxfordshire, is a church chest rudely formed out of the trunk of a tree, with a lid two inches in thickness. The church of Bradford Abbas, Dorsetshire, and Long Sutton, Somersetshire, have also chests thus rudely constructed, to which it is, for want of decorative detail, difficult to fix an approximate date. To other chests covered with panel-work or carved tracery an approximate date may more easily be assigned from the design and detail. In Climping church, Sussex, is an early chest of the thirteenth century, the front of which exhibits a series of plain pointed arches rudely trefoiled in the head, and other carved detail of that age. In Haconby church, Lincolnshire, and Chevington church, Suffolk, are very rich chests covered with carved tracery and detail in the Decorated style of the fourteenth century. In Huttoft church, Lincolnshire, is a chest of the fourteenth century; the front is divided into five compartments, each filled with flowing tracery under a crocketted pedimental canopy. At Brancepeth, Durham, is a chest also of this description. Church chests of the fifteenth century are not uncommon. In Brailes church, Warwickshire, is an ancient chest of this period, covered with panel-work compartments with plain pointed arches foliated in the heads. In St. Peter's church, Saltfleetby, Lincolnshire, is an ancient carved chest of the fifteenth century. In Brixton church, Isle of Wight, is an ancient church chest of the fifteenth century, rudely constructed of boards, but in front ornamented with panel-work in eight compartments, the panels being pointed and trefoiled in the heads, and the spaces between the heads of the panels foliated. In Louth church, Lincolnshire, is a carved chest apparently of the time of Henry VII., embellished with the heads of a King and Queen, the Rose and Crown, and the heraldic supporters to the Royal Arms of the Tudor Period, a Lion and a Dog. In Watford church, Herts, is an old church chest *Temp.* Henry VIII., with the linen fold panel-work in front. In Shanklin church, Isle of Wight, is a chest bearing the date of 1519, on which no architectural ornament is displayed, but the initials T. S. (Thomas Selkstead) are fancifully designed and are separated by the lock, beneath which is a shield with armorial bearings. We sometimes find, more especially in churches bordering or near the eastern coast, a chest of foreign workmanship, termed a "Flanders chest," the carved workmanship on which is sufficient to point out its character, the tracery being flamboyant. The term Flanders chest is to be met with in old inventories. Church chests were sometimes the bequests of individuals; thus A.D. 1392, Nicholas de Scriburn by will left to the church of St. Sampson at York, "*unam cistam de fer ferro ligatam,*" and in the same year Richard de Dalton of York by will as follows: *Item lego summo altari ecclesiæ sanctæ Trinitatis prædictæ (viz.* Ebor) *unam cistam ferro ligatam pro custodiendis ornamentis dicti altaris.*

THE DOMUS INCLUSI, OR HABITABLE CHAMBER FOUND IN MANY OF OUR ANCIENT CHURCHES. Those who have examined, even in the most cursory manner, the internal arrangement of our ancient churches, cannot have failed to notice that many of them contain either over the porch, or over the vestry, or within the tower, or in some other part of the church, a loft or habitable chamber, of which no record or even tradition remains as to the use or purpose for which it was constructed. This loft or chamber, not in all cases, but in many, contains a fire-place and flue, and sometimes also a closet, or jakes, is attached; but this latter is rarely found.

In these days we find this chamber sometimes used as a vestry or a clerical library, containing one of those collections of theological works which a century-and-half ago were, in not a few instances, left by some good pastor for the use of his successors; but frequently this room is now found altogether neglected and disused, or else converted into one of those general receptacles for rubbish, which too often disgrace our churches.

There were religious orders—we can hardly class them as communities—to which, though undoubtedly of early origin, springing up imperceptibly and unheeded, we cannot assign with any degree of precision their actual commencement in the Church. These consisted of individual members of the Church, both of the clergy and of the laity, and of the latter of both sexes, living individually apart and separate from others, some indeed as hermits in wilds and solitudes, far from the abodes of men, and others living amongst the busy haunts of daily life, yet separate and apart,—secluded, as it were, from the world.

The latter were called indifferently *anchorites*, and *recluses*, and lived for the most part, as I hope to show, in a chamber or a cell within or adjoining to a church. This cell was called an *anchorage* or *anker-hold*: in Latin, the *reclusorium* or *domus inclusi*.

The learned Mabillon, in his great and voluminous though unfinished work on the Benedictine annals, which extends, however, only to the twelfth century, its completion having been prevented by his death, alludes, somewhat incidentally, in various parts of that work to the inclusion of anchorites; and under the year 541 of the Christian era, he notices a custom which prevailed in the middle of the sixth century at Vienne, a city in the south of France, where one of the most holy men within that city, addicted to a contemplative life, was shut up in a private cell, that he might continually offer vows to God. Amongst these, he instances the Abbot Leonianus, who, included within the confines of that particular cell, discharged this duty for many years. He also mentions one Theuderius, who, included in the church of S. Laurence, within that city, passed the latter part of his life, during the space of twelve years, in prayer and great austerity, his only means of communication with those who came to him for religious advice being through a little window.*

* Vienna. In eadem vero urbe consuetudo ferebat, unum e sanctioris

The same writer, under the year 589, tells us of certain female recluses who dwelt in the Church of the Holy Cross of the monastery or nunnery of S. Radegund, from whom the *rule* observed in their seclusion might be learned. From this account it appears that a cell having been prepared, and an intended "*ankress*" having requested permission to inhabit it, she was, after certain ceremonies, included or shut up, and the door through which she entered was then blocked up.⁕

A rule similar to this is said to have prevailed with Grimlaic. No entrance was left by which admission into the cell could be obtained after the anchorite was once enclosed; but the cell was constructed like a tower, and only a window left through which the recluse could hold intercourse with those without, and through which sustenance might be received. The bishop presided at

vitæ hominibus intra urbem recludi in privata cellula qui uni vitæ contemplativæ addictus, vota populi Deo assidue offeret. Fuerat ex eo numero Leonianus Abbas superius laudatus, qui Viennæ claustro peculiaris cellæ conclusus per multos annos hoc officio functus est. . . . . . Electus ad id muneris a Philippo Theuderius, Severiano præposito postremi hujus monasterii curam commisit, atque in monte Quirinali intra urbem ad basilicam sancti Laurentii reclusus, reliquum vitæ nimirum per duodecim annos, in oratione et magnâ distrinctione exegit, ita ut advenientibus monita salutis nou nisi per modicam fenestram daret.—*Mabillon, Annales Benedictini, sub anno* 541.

⁕ Non licet pretermittere hoc loco, reclusas exstitisse quondam in parthenone Sanctæ Crucis, uti Gregorius tradit, ex quo etiam discimus ritum, qui in earum reclusione servabatur. Is quippe agens de quadam sanctimoniali, quæ viso ad compunctionem animata fuerat, subdit eam post dies paucos rogavisse abbatissam ut sibi in qua recluderetur, cellulam præperaret. Tum cellulâ quantocius perfectâ, puellam petisse ut recludi permitteretur, quod cum ei permissum fuerit, congregatis virginibus cum magno psallentio, accensis lampadibus, tenente sibi beatâ Radegunde manum, ad locum usque perducitur. Et sic valefaciens omnibus, et osculans singulas quasque reclusa est; obstructoque aditu per quem ingressa fuerat ibi nunc orationi et lectioni vacat, ait Gregorius.—*Ibid, sub anno* 589.

the shutting up of the recluse, and sealed the door of the cell after the seclusion of the votary.ʳ

In the year 741 we find that the rule pertaining to recluses was not so strict, but that they were occasionally permitted to go abroad. This, however, was only allowed on special occasions, as when Sigibert, a recluse and monk of S. Denis, near Paris, was sent by Charles, King of the Franks, with Grimo, Abbot of Corbie, on a mission to Pope Gregory.ˢ

In the year 793 we find that one Alfrida lived as a recluse in a cell on the south of the church of Croyland, near the high altar.ᵗ

The same learned writer tells us, under the year 916, that this was a noted practice amongst both sexes,—of becoming recluses; and that the first who prescribed a rule for recluses was Grimlaic, a priest, although the custom of seclusion prevailed before his time.ᵘ

---

ʳ Idem fere ritus legitur apud Grimlaicum in regula solitariorum. Sic apud eundem Gregorium de Hospitio recluso apud Niciam legitur, nullam patuisse in ejus cellulam aditum, quæ instar turris fabricata erat, solamque relictam fuisse fenestram per quam scilicet vitæ necessaria accipiebat, aut exterius loquebatur. Grimlaicus addit, Episcopum reclusioni præfuisse et ostium retrusionis cellulæ aposphragismo, id est sigillo, suo solitum sigillare.—*Ibid.*

ˢ Porro ex hac Sigoberti reclusi legatione colligimus, reclusos non usque adeo reclusionis legibus fuisse adstrictos, ut aliquando sibi ex cella sui egredi, saltem ob publicam utilitatem, non liceret.—*Mabillon, Annales Benedictini, sub anno* 741.

ᵗ Ælfrida, vero aliis Ælthrida, quam Ingulfus Etheldritam vocat, in australi parte ecclesiæ crulandensis contra majus altare cellâ reclusa vixit. —*Ibid, sub anno* 793.

ᵘ Celebre quondam fuit apud veteres monachos, atque adeo apud nostros reclusorum utriusque sexus institutum. . . . . Primus regulam ejusmodi reclusis præscripsit Grimlaicus presbyter, qui reclusionis modum et formam, qualis sine dubio ante ipsum usitata erat, describit.—*Ibid, sub anno* 916.

Recluses sometimes dwelt in monasteries, in cells set apart for that purpose.* The rule of S. Benedict, as applied to these recluses, appears to have been most general. In the eleventh century the cell thus set apart was called the "reclusorium."†

Under the year 1094, mention is made of Hildeburgis, a noble matron, who betook herself to the monastery of S. Martin, where, having taken upon her the religious habit, she caused a small habitation to be constructed for her adjoining the north side of the church, and there passed the remainder of her life in great austerity, employing herself in working ornaments and vestments for the church. Here she died, full of years and good works, in the year 1115, and was buried in the church of S. Martin, near to the wall of her cell.‡

Doubtless, in the great work of Mabillon,—from which the foregoing notices are deduced, though not in so ample a manner as he has given them,—much more on recluses would have appeared, but for the death of this learned and indefatigable Benedictine leaving this work

---

* In monasteriis virorum non modo reclusi erant, sed etiam reclusæ in cellis a monasterio separatis admittebantur.—*Ibid.*

† Moris erat tunc temporis ut religiosæ feminæ prope monasteria nostra recluderentur, ut in illo *reclusorio*, vitam a sæculi corruptelis longe remotam ducerent.—*Ibid, sub anno* 1033.

‡ Hildeburgis, nobilis matrona, &c. . . . . . . . Postremo ad Pontisarense sancti Martini cœnobium se recepit ubi sanctimonialis habitu suscepto a Theobaldo abbate, exiguam sibi domunculam ad septentrionalem ecclesiæ partem fabricari curavit, ibique reliquum vitæ in magna districtione exegit, ornamenta ecclesiæ, fratribusque vestimenta conficere solita. Infirmorum quoque domum suis impensis, necnon pauperum xenodochium ædificavit. Ibidem piissima illa matrona, plena dierum et operum bonorum, decessit iii nonas Junii sub annum 1115 sepulta in Sancti Martini ecclesia, ad murum cellulæ suæ adhærentem.—*Ibid, sub anno* 1094.

unfinished, it having been carried on only to the latter part of the twelfth century.

We find in the Statutes of the Synod held by Richard, Bishop of Chichester, A.D. 1246, one respecting recluses. By this they were enjoined not to admit or have any person in their dwellings of whom grave suspicion might arise. Their windows were also required to be narrow and convenient; they were permitted to have intercourse with those persons only whose character and conduct did not admit of suspicion. The custody of the vestments of the church was not, except in cases of necessity, to be delivered to female recluses.[a]

The religious office for the including of anchorites, "Reclusio Anachoritarum," we find in the Pontifical—written in the fourteenth century—known as that of Lacy, Bishop of Exeter, in the fifteenth century. In this service the sacrament of extreme unction was administered, and the prayer of commendation for the soul of the recluse was offered, lest, being prevented by death, he should stand in need of those rites of the Church. Part of the funeral service was also performed, and the "domus," "reclusorium," or anchorage, is represented as a sepulchre into which the recluse entered, being, as it were, thenceforth dead to the world.

In the manual according to the Sarum use, *Manuale*

[a] De inclusis.—Inclusis etiam præcipimus, ne quam personam in domibus suis recipiant vel habeant, de qua sinistra suspicio oriatur. Fenestras quoque arctas habeant et honestas; eisdem etiam cum his tantummodo personis secretum tractatum habere permittemus, quarum gravitas et honestas suspicionem non admittit. Inclusis vero mulieribus custodia vestimentorum ecclesiæ non tradatur, quodsi necessitas hoc exegerit, ita caute, tradi mandamus, ut non inspiciantur inclusæ a tradente.—Statuta Synodalia Ricardi Cicestren episcopi A.D. 1246.—*Wilkins' Concilia.*

*ad usum insignis ecclesie Sarum,* published at Paris, A.D. 1515,[b] this office, headed "Servitium Includendorum," is essentially the same as in the Pontifical of Bishop Lacy, differing only in particulars; but in the Pontifical there is another service, headed "Benedictio Heremitarum," which does not appear in the manual. The profession of a hermit and an anchorite, or recluse, ought not to be confounded together.

Helyot, in his voluminous work, *Histoire des Ordres Religieux,* has omitted all notice of the anchorites or recluses,—at least, I cannot find any part of his work where he has treated of them,—and information respecting them, their residences, and mode of living, is only to be drawn from scattered sources. It is probable, however, that considerable information, both on this and many other points of interest relative to the ancient discipline of the Church, might be obtained from an inspection and examination of the bishops' registers in different dioceses, were such rendered readily accessible.

From an instrument in one of these registers, it appears that adjoining the chapel of Bablake, at Coventry, now better known as S. John's church, was a place built for the including of an anchorite. This is called in the License[c] of Roger, Bishop of Coventry, dated in 1362,

[b] I have been favoured with the loan of a copy of this scarce and valuable work, which is preserved in a highly interesting collection of early printed works and manuscripts, which formed the library of a clergyman at or soon after the Reformation, and which collection is now in the possession of Henry Sherbrooke, Esq., of Oxton, Nottinghamshire.

[c] Rogerus &c., dilecto in Christo filio Roberto de Worthin Cap. salutem, &c. Precipue devotionis affectum, quem ad serviendum Deo in *reclusorio* juxta Capellam Sancti Joh. Baptiste in civitate Coventriensi constructo, et spretis mundi deliciis et ipsius vagis discurribus contemptis, habere te

a "reclusorium;" no traces, however, of this annexed building at present remain.

Blomefield, in his *History of Norwich*, has preserved many particulars respecting "ankers" and "ankresses," who dwelt in that city, some of which particulars I shall proceed to notice.

In the east part of the churchyard of S. Julian's church, Norwich, stood an anchorage, in which an ankeress, or recluse, dwelt till the Dissolution, when the house was demolished, though Blomefield informs us that foundations might be seen in his time. In 1393, Lady Julian, the ankeress here, was a strict recluse, and had two servants to attend her in her old age. Anno 1443, this woman was in those days esteemed one of the greatest holiness. In 1472, Dame Agnes was recluse here. In 1481, Dame Elizabeth Scott. In 1510, Lady Elizabeth. In 1524, Dame Agnes Edrygge.

There was very anciently an *anchorage* in the churchyard of S. Etheldred's church, Norwich, which was rebuilt A.D. 1305, where an *anchor* continually resided till the Reformation, when it was pulled down, and the Grange, or Tithe Barn, at Brakendale, was built with its timber.

Joining to the north side of S. Edward's church, in the same city, was a cell, the ruins of which might be

asseres, propensius intuentes, ac volentes te, consideratione nobilis domine, domine Isabelle Regine Anglie nobis pro te supplicante in hujus laudabili proposito confovere, ut in prefato reclusorio morari possis, et recludi et vitam tuam in eodem ducere in tui laudibus Redemptoris, licentiam tibi quantum in nobis est concedi per presentes, quibus sigillum nostrum duximus apponendum. Dat apud Heywood, 5 Kal. Dec. A.D. MCCCLXII. et Consecrationis nostræ tricessimo sexto.—*Dugdale's Antiq. Warwickshire*, 2nd edition, p. 193.

seen so late as A.D. 1744, in which a recluse continually dwelt, and most persons that died in that city left small legacies towards her support. In 1428, Lady Joan was anchoress here, to whom Walter Sedman left xxs., and xLd. to each of her servants. In 1458, Dame Anneys, or Agnes Kyte, was recluse here. In 1516, Margaret Norman, widow, was buried here, and gave a legacy to the Lady Anchoress by the church.

The church of S. John the Evangelist, in Southgate, Norwich, was, about the year 1300, annexed to the parish of S. Peter per Montergate; it was then purchased by the Grey Friars to augment their site, when the whole was pulled down, except a small part left for an *anchorage*, in which they placed an *anker*, to whom part of the churchyard was assigned for a garden.

Anciently, there was a recluse dwelt in a little cell joining to the north side of the steeple of the church of S. John the Baptist, of Timberhill, Norwich, but it was down before the Dissolution.

There was also, anciently, an anchor, or hermit, who had an anchorage in or adjoining to the church of All Saints, in the same city.

In Henry III.'s time, there was a recluse who dwelt in the churchyard of the church of S. John the Baptist and the Holy Sepulchre, in Ber-street, Norwich.

In the monastery of the Carmelites, or White Friars, Norwich, there were two *anchorages*, or *anker-houses*, one for a man, who was admitted Brother of the House, and the other for a woman, who was also admitted Sister thereof,—the last under the Chapel of the Holy Cross, which, in Blomefield's time, was still standing, though converted into dwelling houses; the former stood by S. Martin's Bridge, on the east side of the street, and a small garden belonging to it joined to the river. In 1442, December 2, the Lady Emma, recluse, or anchoress, and religious Sister of the Carmelites Order, was buried in their church; and in 1443, Thomas Scroop was anchorite in this house. In 1465, Brother John Castleacre, a priest, was anchorite. In 1494, there were legacies given to the anchor of the White Friars. Thomas Scroop, originally a Benedictine monk, in 1430, took the habit of a Carmelite Friar, and led an anchorite's life here for many years, seldom going out of his cell but when he preached; but about the year 1446, Pope Eugenius IV. made him Bishop of Down, in Ireland, which he afterwards resigned, and came again to his convent, and became suffragan to the Bishop of Norwich. He died at Lowestoft in 1491, and was there buried, being near a hundred years old.

Well might Blomefield observe, that "there were many of these anchorets and anchoresses in the city of Norwich."

Henry de Knyghton, a canon of Leicester, in his chronicle *De Eventibus Angliæ*, states, that in the year 1392, Courteney, Archbishop of Canterbury, visited the diocese of Lincoln, and in his visitation he came to the Abbey of Leicester, and there in full chapter confirmed sentence of excommunication against the Lollards, or Wyclyffites, and against all who held or entertained, or might thereafter hold or entertain the errors and opinions of Master John Wycliffe in the diocese of Lincoln. And on the morrow, which was the day of All Souls, the archbishop fulminated sentence of excommunication, with the Cross erect, candles burning and bells ringing according to wont, on nine persons of the town of Leicester; and about evensong the archbishop went to the church of S. Peter, to a certain anchoress named Matilda, who dwelt there as a recluse, and arguing with her on the errors and opinions of the Lollards, which it would seem she entertained, he cited her to appear before him on the following Sabbath, at the Abbey of St. James, Northampton, which having done, and having confessed her errors, and penance having been enjoined her, she again entered her anchorage, or *reclusorium*.

The same writer, under the year 1382, gives an account of a priest then dwelling at Leicester, one William de Swyndurby, commonly called William the Hermit, who, having a character for sanctity, the canons of Leicester received to lodge in a certain chamber within the church, "*in quâdam camerá infra ecclesiam*," that they might procure him sustenance, together with a pension, after the manner of other priests.

In that one of the cycle of the numerous romances of the life and death and adventures of King Arthur, renowned in British story, and the Knights of the Round Table, most familiar to us—having been translated or composed from earlier romances on the same subject by Sir Thomas Malory, Knight, in the reign of Edward IV., and entitled *La Mort d'Arthur,*—we have some curious incidental references to customs and practices common in the fifteenth century; amongst which is the following notice of a recluse, in a relation of the adventures of Sir

Launcelot :—" Then he armed him, and took his horse, and as he rode that way he saw a chapel where was a recluse, which had a window that she might see up to the altar; and all aloud she called Sir Launcelot, because he seemed a knight errant; and then he came, and she asked him what he was, and of what place, and what he seeked."

That anchorites were numerous we find from bequests to several specified in the will of Henry, third Lord Scrope, of Masham, dated the 23rd of June, 1415. By this was bequeathed to John, the anchorite of Westminster, cs., and the pair of beads which the testator was accustomed to use; to Robert, the recluse (recluso) of Beverley, xls.; to a certain chaplain dwelling in York, in a street called Gilligate, in the church of S. Mary, viijs. ivd.; to Thomas, the chaplain, dwelling (commoranti continuo) in the church of S. Nicholas, Gloucester, xiijs. ivd.; to the anchoret of Stafford, xiijs. ivd.; of Kurkebiske, xiijs. ivd.; of Wath, xxs.; of Peesholme, near York, xiijs. ivd.; to Elizabeth, late servant of the anchoret at Hampole, ———; to the recluse at Newcastle, in the house of the Dominicans, xiijs. ivd.; to the recluse at Kenby Ferry, xiijs. ivd. To the several anchorets of Wigton, of Castre, of Thorganby near Colyngwith, of Leek near Upsale, of Gainsburg, of Kneesall near Southwell, of Staunford, living in the parish church there, of Dertford—each xiijs. ivd.; also to every anchoret and recluse dwelling in London or its suburbs, vis. viijd.; also to every anchoret and recluse dwelling in York and its suburbs (except such as were already named), vis. viijd.; to the anchoret of Shrewsbury, at the Dominican convent there, xxs.; also to every other anchoret and anchoritess that could be easily found within three months after his decease, vis. viij*d*.[d]

We are left much in the dark as to the destiny of the anchorites on the general suppression of the religious houses and chantries. They appear to have been dependent on casual charity, or their own resources, rather than on endowments of a permanent nature. Whether at the eventful era of the Reformation they were still suffered to remain in their seclusion until death; or whether they had again to encounter the gaze and temptations of the world, from which they had voluntarily seceded; or whether they were at liberty to adopt either course, is a matter involved in considerable obscurity. From what, however, Becon states of them in his *Reliques of Rome*, published A.D. 1563, we are led to the inference that this class of individual recluses was not at that time extinct, since he treats of them not in the past but in the present tense. But his account shall speak for itself:—

"As touching the monastical sect of recluses, and such as be shutte up within walles, there unto death continuall to remayne, giving themselves to the mortification of carnal effectes, to the contemplation of heavenly and spirituall thinges, to abstinence, to prayer, and to such other ghostly exercises as men dead to the worlde, and havyng their lyfe hidden with Christ, I have not to write: forasmuch as I cannot hitherto fynde, probably in any author, whence the profession of anckers and ankresses had the begennyng and foundation, although

[d] Scrope's *History of Castle Combe*.

in this behalf I have talked with men of that profession, which could very little or nothing say of the matter. Notwithstanding as the whyte fryers father that order on Helias the prophet (but falsly), so likewise do the ankers and ankresses make that holy and virtuous matrone, Judith, their patroness and foundresse. But how unaptly who seeth not? Their profession and religion diffreth as far from the maners of Judith, as light from darknesse, or God from the devill, as it shall manefestly appere to them that will diligentlye conferre the history of Judith with their life and conversation. Judith made herself a privy chamber where she dwelt (sayth the Scripture), being closed in with her maydens. Our recluses also close themselves within the walles, but they suffer no man to be there with them. Judith ware a smocke of heare, but our recluses are both softly and finely apparaled. Judith fasted all the days of her lyfe, few excepted. Our recluses eate and drinke at all tymes of the beste, being of the number of them *Qui curios simulant et Bacchanalia vivunt.* Judith was a woman of a very good report. Our recluses are reported to be superstitious and idolatrous persons, and such as all good men flye their company. Judith feared the Lord greatly, and lyved according to His holy word. Our recluses fear the pope, and gladly doe what his pleasure is to command them. Judith lyved of her own substance and goods, putting no man to charge. Our recluses, as persons only borne to consume the good fruites of the erth, lyve idely of the labour of other men's handes. Judith, when tyme required, came out of her closet to do good unto other. Our recluses never came out of their lobbeies, sincke or swimme the people. Judith put herself in jeopardy for to do good to the commune countrye. Our recluses are unprofitable cloddes of the earth, doing good to no man. Who seeth not now, how farre our ankers and ankresses differe from the manners and life of this vertuous and godly woman, Judith, so that they cannot justly claime her to be their patronesse? Of some idle and superstitious heremite, borrowed they their idle and superstitious religion. For who knoweth not that our recluses have grates of yron in theyr spelurckes and dennes, out of the which they looke, as owles out of an yvye todde, when they will vouchesafe to speake with any man at whose hand they hope for advantage? So reade we in Vitis Patrum, that John the Heremite so inclosed himself in his heremitage, that no person came in unto him, to them that came to visite him, he spake thorow a windowe onely. Our ankers and ankresses professe nothing but a solitary lyfe in their hallowed house wherein they are inclosed, wyth the vowe of obedience to the pope, and to their ordinary bishop. Their apparell is indifferent, so it be dissonant from the laity. No kind of meates they are forbidden to eat. At midnight they are bound to say certain praiers. Their profession is counted to be among all other professions so hardye and so streight, that they may by no means be suffered to come out of their houses."

Such being the particular class of devotees who appear to have dwelt in the habitable chambers or cells in our churches, a brief enumeration and description of a few of these cells or chambers will follow, though a more extended list and fuller particulars may readily be collected and obtained.

In the church of Daglingworth, in Gloucestershire, a structure curious for the admixture of Anglo-Saxon and Norman detail, traces exist of a loft or chamber over the western part of the nave, but now thrown open to it by the removal of the flooring; and in the east wall of this chamber is a stone table or altar, not easily accessible, but the details of which make it to be of as early a date as the twelfth century. No anchor-hold, or *reclusorium*, now exists in, or adjoining to, S. John's church, Coventry, to which I have formerly adverted as containing one, but this church, which is not parochial, has undergone many vicissitudes since the Reformation. The church of S. Peter, at Leicester, to which historical allusion has been made, was long since demolished; but in the church of S. Mary, near the castle, in that town, there was, until within the last few years, a loft, or habitable chamber, at the west end of the north aisle, with a window at the east end, looking up the aisle, and fully answering the description of Knyghton, as being "Camera infra ecclesiam." Over the south porch of Southwell Minster, is a chamber, or loft, with a fireplace; this is a structure of the twelfth century.

Adjoining the little mountain church of S. Patricio, about five miles from Crickhowel, South Wales, is an attached building or cell, answering to that of the recluse described in *La Mort d'Arthur*. It contains on the east side a stone altar, above which is a small window, now blocked up, which looked towards the altar, but there was no other internal communication between this cell and the church, to the west end of which it is annexed. It appears as if destined for a recluse who was also a priest. In this interesting little church, I may remark that the rood-loft is still existing in a more perfect state than we usually find it; and beneath the rood-loft are two ancient stone altars, the only remaining instances of the rood-loft altars I have hitherto met with. In this little church, then, consisting of a nave and chancel only, with the adjoining cell at the west end, are three ancient stone altars, a greater number than I have ever met with as still existing in any of our ancient churches, with the exception of Arundel church, Sussex, in which there are four.

The north transept of Clifton Campville church, Staffordshire, a structure of the fourteenth century, is vaulted and groined with stone: it measures 17 feet from north to south, and 12 feet from east to west. Over this is a loft or chamber, apparently an anchor-hold, or *domus inclusi*, access to which is obtained by means of a newel staircase in the south-east angle, from a doorway at the north-east angle of the chancel. A small window on the south side of this chamber, now blocked up, afforded a view into the interior of the church. The roof of this chamber has been lowered, and all the windows blocked up. The tower of Boyton church, Wiltshire, is on the north side of the church; and adjoining the tower on the west side, and communicating with it is a room which appears to have been once permanently inhabited; and in the north-east angle of this room is a fire-place. In the tower of Upton church, Nottinghamshire, there is also a room with a fire-place.

On the north side of the chancel of Chipping Norton church, Oxfordshire, is a revestry, which still contains an ancient stone altar, with its appurtenances—*viz.*, a piscina in the wall on the south side, and a bracket for an image projecting from the east wall, north of the altar. Over this revestry is a loft, or chamber, to which access is obtained by means of a staircase in the north-west angle. Apertures in the walls enabled the recluse, probably a priest, here dwelling, to overlook the chancel and north aisle of the church.

Adjoining the north side of the chancel of Warmington church, Warwickshire, is a revestry, entered through an ogee-headed doorway in the north wall of the chancel, down a descent of three steps. This revestry contains an ancient stone altar projecting from a square-headed window in the east wall; and near the altar, in the same wall, is a piscina. In the south-west angle of this revestry is a flight of stone steps, leading up to a chamber, or loft. This chamber contains in the west wall a fire-place; in the north-west angle a retiring closet, or jakes ; and in the south wall a small Pointed window of Decorated character, through which the high altar in the chancel might be viewed. In the north wall there appears to have been a Pointed window filled with decorated tracery; and in the east wall is another Decorated window. This is one of the most interesting and complete specimens of the *domus inclusi* I have met with. In S. Kenelm's chapel, near Hales Owen, Salop, was formerly a loft, or chamber, with a closet, or jakes, contained in or contiguous to it. This has been destroyed.

Rooms over Porches. Over the south porch, Wigenhale church, Norfolk, is a room, access to which is by a turret staircase west of the door. Outwell church, Norfolk, has a groined south porch with a room over, access to which is obtained by a turret staircase west of the door. The south porch of Walpole S. Peter's, Norfolk, is groined in two divisions; in the north-west angle of the porch, and opening into it, is a newell staircase to a room above. The south door of Uffington church, Berkshire, has a fine Early English porch, with a groined vault; over this is a room which contains an original fire-place and chimney. At Mettingham church, Suffolk, is a flue in the porch, with an aperture for a fire-grate or cradle. At Cromer, in Norfolk, the north porch, a structure of the fifteenth century, now in ruins, formerly had a chamber or room over it, entered from the south-west angle, up a turret through the church. This room had a fire-place, constructed of brick-work, in the south wall; eastward of this was a small retiring closet, probably used as a jakes. Over the south porch of Grantham church, Lincolnshire, is a chamber or ankorage, with a fire-place in the south-west corner, and a window projecting into the church. This room now contains a library of old, and, apparently, valuable theological works. The south porch of Nantwich church, Cheshire, has a room over containing a fire-place. This is of the fifteenth century. The south porch of Martham church, Norfolk, has a room over. The south porch of Worstead church, Norfolk, is groined, and has a room over. Over the south porch of S. Margaret's church, Lowestoft, Suffolk, is a room, called the maid's chamber; this had formerly a fire-place; the entrance is within the church, in the south wall, west of the doorway. At Wrotham church, Kent, is a *domus*, or habitable chamber, over the porch, with a fire-place on the west side. Marldon church, Devon, has a chamber over the south porch, 8 feet 6 inches by 7 feet 9 inches; this room contains a fire-place; the porch beneath is groined, and of the fifteenth century.

IN TOWERS. In the tower of Buckminster church, Leicestershire, is a fire-place. The tower of Brailes church, Warwickshire, also contains a fire-place; this is on the first floor. In the tower of Bradeston church, Norfolk, are the remains of a fire-place and funnel in the basement story; the tube is carried to the height of about 8 feet, without any external or internal projection from the otherwise solid rubble walls; and the aperture on the north face is without the slightest embellishment. In the tower of Ranworth church, Norfolk, is a similar formed hearth, but the flue is not visible. In the circular tower of Thorp Abbots church, Norfolk, on the north side of the basement story is a fire-place and chimney, the flue of which runs up the wall nine inches square, the smoke escaping from a small north loop. The tower of Battlefield church, Salop, has on the second floor a fire-place within the thickness of the wall, with an opening to let off the smoke outside of the western window of the bell chamber.

OVER VESTRIES. On the north side of the chancel of Tenbury church, Worcestershire, is a vestry-room entered through a Decorated doorway; over this room is a chamber which has a window (now blocked up) which opened into the chancel. Attached to the north side of the chancel of Dartford church, Kent, is an oblong chamber with a slit window opening into the chancel. On the east side of this chamber, which is a structure of the fifteenth century, is a fire-place. The vestry on the north side of the chancel of Cropredy church has a chamber above it. The vestry of Edgecote church, Oxfordshire, on the north side of the chancel, has a room over it. Above the vestry, on the north side of the chancel of S. Margaret's church, Leicester, is a chamber, access to which is by a turret staircase at the southwest angle; this is a structure of the fifteenth century. Horsham church, Sussex, has a chamber over the vestry. At Paignton church, Devon, over the vestry, which is on the *south* side of the chancel, is a room containing a fireplace in the south wall, and an original door. This room is 18 feet 6 inches long by 9 feet 8 inches wide.

In the regulations for recluses in Bavaria, the cell adjoining or forming part of a church, and described as *Inclusa, id est domus inclusi*, was required to be constructed of stone, twelve feet square, with three windows, of which one was towards the chancel or choir, through which the Host might be received in partaking of the Eucharistic Sacrifice; another opposite to the former, opening through the external wall of the church, by means of which provisions necessary to sustain life might be received; and a third closed with glass or horn, for the admission of light.[c]

[c] In ordine Inclusorum apud Raderum in Bavaria sancta cellula sic de-

But the small rooms over the porches of our churches are far more numerous than those over the vestries or elsewhere, and many of these have been additions, and constructed at a period subsequent to that of the substructures on which they have been raised. These lofts over porches were, I imagine, inhabited by lay recluses, male or female.

A tradition respecting a female recluse inhabiting one of these cells is related by Dickenson, in his *History of Southwell*. In describing the porch of the chapel of Holme, he says—" Over this porch is a chamber, called as far back as memory or tradition reach, Nan Scott's Chamber. The story of which this lady is the heroine, has been handed down with a degree of precision and uniformity which entitles it to more credit than most such tales deserve. The last great plague which visited this kingdom is reported to have made particular havoc in the village of Holme; which is likely enough to have happened, from its vicinity to Newark, where it is known to have raged with peculiar violence. During its influence, a woman of the name of Ann Scott is said to have retired to this chamber with a sufficient quantity of food to serve her for several weeks. Having remained there unmolested till her provisions were exhausted, she came from her hiding-place either to procure more or to return to her former habitation, as circumstances might direct her choice. To her great surprise, she found the village entirely deserted, only

---

scribitur:—Inclusa, id est domus inclusi, debet esse lapidea, longitudo et latitudo in 12 pedes, habeat tres fenestras, unam contra chorum per quam corpus Christi accipiat, alteram in opposito per quam victum recipiat tertiam unde lucem habeat quæ semper debet esse clausa vitro vel cornu.

one person of its former inhabitants, except herself, being then alive. Attached to this asylum, and shocked by the horrors of the scene without, she is said to have returned to her retreat, and to have continued in it till her death, at an advanced period of life. A few years since many of her habiliments were remaining in this chamber, as also a table, the size of which evidently manifested it to have been constructed within the room, with some smaller pieces of furniture."

Blomefield, the historian of Norfolk, in his account of Gessing, makes mention of one of the rectors, John Gibbs, A.M., who was presented by King Charles the Second, A.D. 1668, and continued as rector till A.D. 1690, when he was rejected as a nonjuror. "He was," says the historian, "an odd but harmless man, both in life and conversation. After his ejection, he dwelt in the north-porch chamber, and laid on the stairs that led up to the rood-loft between the church and chancel, having a window at his head, so that he could lie in his narrow couch and see the altar. He lived to be very old, and at his death was buried at Frenze."

Such was the last of the recluses, if so he may be called, of whom I can find any notice.

THE CAPELLÆ, CARNARIÆ, CHARNEL CHAPELS, AND CHARNEL VAULTS. Beneath certain of our churches, crypts and vaults may be found, either still containing, or which have heretofore contained, a mass of human sculls and bones dug up from the cemeteries surrounding those churches. From the exhibition of these remains, wild and unfounded tales have originated, pointing erroneously to some supposed battles fought in the neighbourhood, of which no historical notices exist, whilst mere tradition is ofttimes founded upon ignorance. The real fact being that these accumulations of bones were but the results of a common practice formerly prevalent, of which I now attempt to treat in elucidation.

The Author, whoever he was, of *A Description or Breife Declaration of all the Ancient Monuments, Rites, and Customes belonginge or beinge within the Monastical Church of Durham before the Suppression*, written in 1593, observes, of the centrie or cemetery garth of that monastery, "Att the easte end of the said chapter-howse there is a garth called the centrie garth, where all the priors and monnckes was buried. In the said garth there was a vaulte all sett, within either syde with maison wourke of stone, and likewise at eyther end, and over the myddes of the said vaut, there dyd ly a faire through stone, and at either side of the stone was open, so that when any of the monnckes was buryed, looke what bones was in his grave, they were taiken when he was buryed and throwne in the said vaulte which vaut was made for the same purpose to be a charnell to cast dead men's bones in."

In a work published in Paris in 1666, entitled *Le Parfaict Ecclesiastique, &c.*, by Claude de la Croix, a priest of the Gallican Church, being in fact a clergyman's *vade mecum*, certain queries are set forth for the guidance of a priest on the visitation of his parish, and amongst those relative to the grave-yard or burial-ground. "*Du Cimetiere*" one, the ninth query is,—"*S'il y a un retranchment couvert en un coin, pour y ranger les ossemens des trespassez, en cas qu'il y en eust en abondance ?*" "Is there a bye-place covered over in a corner wherein to deposit the bones of the dead in case they should be in great numbers?"

Dom Claude de Vert, a learned Ritualistic writer of the Gallican Church, in his work, one of deserved repute, *Explication simple, Litterale et Historique des ceremonies de L'Eglise*, published in 1713, under the word *Charnier*, observes in a note, "*C'est à dire le lieu qui est auprés, ou autour des Eglises ou des Cimitieres, ou l'on enterre les trespassez, et ou l'on met aussi quelques fois des restes et des ossemens entrumez.*"

In that interesting work, the *Voyage Litteraire*, an account of the travels of two of the learned Benedictines of S. Maur, Martene and Durand, to 800 Abbeys in France, with the view of obtaining materials for the second edition of the *Gallia Christiana*, and the *Thesaurus novus Anecdotorum*, we find in the account given of the celebrated Abbey of Clairvaux, where S. Bernard died and was buried, they describe the chapel of the Counts of Flanders.—"*Sous l'autel de cette chapelle, il y a une belle crypte voutée, dans laquelle sont arrangez les ossemens des religieux qui vivoient du temps de Saint Bernard.*" Under the altar of this chapel is a fair vaulted crypt, in which are disposed the bones of those religious who lived in the time of S. Bernard.

In the close on the north side of Worcester Cathedral the level greensward covers, or did cover, in one part a crypt full of bones, no longer visible to the eye. Of this we find it recorded that William de Blois, Bishop of Worcester from A.D. 1218 to A.D. 1236, in the year

1224 provided a subterraneous vault for the reception of the bones displaced in the cemetery adjoining the cathedral on the preparing of new graves, and over this crypt, which was on the north side of the cathedral near the west end, but a detached building, he raised a chapel called the *Capella Carnaria*, or the chapel of the charnel house. This chapel was consecrated by Bishop Walter de Cantilupe, successor to Bishop Blois, in the thirteenth century. It was amply endowed, and in it masses were celebrated daily for the repose of the souls of those whose bones were placed in the crypt. This chapel, after the Reformation, fell into ruins. It appears to have been demolished about the year 1677: and Green, in his *History of Worcester*, published in 1796, says,—"The only vestiges of the chapel that remain are part of the north and south walls, but the cript which is underneath it remains entire. Its length is fifty-eight feet, its breadth twenty-two, and its height about fourteen. It contains a vast quantity of bones, which, although now in some disorder, seem to have been curiously assorted, and piled up in two rows along its sides, leaving a passage between them from its west entrance to its east end."

Another instance of a charnel house, with the *Capella Carnaria* or chapel over, was formerly on the north side of Pardon churchyard, on the north of old S. Paul's Cathedral, London. Here was a large charnel house for the bones of the dead, and over it a chapel, built about the year 1282. In the reign of Richard the Second, this chapel had got into great decay, and attention being called to it, it was put into a proper state of reparation. A brotherhood, called the Fraternity of All Souls, was founded in this chapel upon the charnel in the year 1379. This chapel was pulled down in the year 1549, and the bones of the dead, couched up in a charnel under the chapel, were conveyed from thence into Finsbury Field (by the report of him who paid for the carriage), amounting to more than one thousand cart-loads, and there laid in a moorish ground, in short space after raised by soilage of the city upon them to bear three windmills. The chapel and charnel were converted into dwelling-houses, warehouses, and sheds, for stationers built before it.

At Bury St. Edmunds, the churchyard, anciently known as the cemetery of S. Edmund, still contains some ivy-covered walls, being the remains of the chapel of the charnel founded by Abbot John de Northwold, A.D. 1301. The Charter of Foundation states that "Lately passing over the cemetery allotted for the burial of the common people," the Abbot had observed " not without sorrow of heart, and pressure of vehement grief," how very many of the graves had been violated by the multiplied burial of bodies, and the bones of the buried "indecently cast forth and left." He therefore directed a chapel to be built, "covered with stone competently, under the cavity of which the buried bones may be laid up or buried reverentially and decently in future," and that "the place shall happily be rendered most famous by the perpetual celebration of the masses of two chaplains."*

Blomefield, in his history of Norwich, notices the charnel house westward of the cathedral, and states

---

*Timm's Handbook to Bury St. Edmunds.*

that it was founded by John Salmon, Bishop of Norwich, who died A.D. 1325. In Blomefield's time the upper charnel chapel, dedicated in honour of S. John the Evangelist, was the Free School. Underneath was the lower charnel chapel and charnel house itself, then used for a vault or cellar. This chapel was also dedicated in honour of the same Saint, and here the keeper of the lower charnel officiated daily, as they all did, *viz.*, the chaplains, four in number, of the upper chapel, for the souls of Salmon, his, the founder's, father, Amy, his mother, his own soul, and those of all the departed Bishops of Norwich, in particular; all the dead in general, and in particular for the souls of all those whose bones were reposited in the vault of this charnel, in which, with the leave of the Sacrist who kept the key of the vault, the bones of all such as were buried in Norwich might be brought into it, if dry and clean from flesh, there to be decently reserved till the last day. "Whether," says Blomefield, "the bones were piled in good order, the sculls, arms, and leg bones in their distinct rows and courses, as in many charnel houses, I cannot say, nor how they were disposed of, when removed after the Reformation."

Blomefield describes the charnel as "an arched vault, supported by two rows of pillars, 14 feet high; at the entrance, on the right hand, was a holy water stone, (*i.e.*, a stoup), and on the other side a niche, where formerly an image stood."

This building is still standing, and is, perhaps, the most perfect specimen we have existing of the *Capella Carnaria*.

The same historian, Blomefield, informs us that the vault under the chancel of S. Gregory's church, in the same city of Norwich, was a charnel.

Under a portion of Ripon Cathedral is a Norman crypt, which formerly contained a large quantity of human sculls and bones; whether these have been removed, I know not.

A similar collection was a few years ago to be found in a crypt of the fourteenth century under the south aisle of Tamworth church, Staffordshire.

A crypt beneath the Abbey church, Waltham, is also said to have contained a large quantity of human bones.

In some repairs done within the church of S. Michael, at Oxford, about twenty-seven years ago, the workmen discovered a vault not previously known to exist, which formed a receptacle for a large quantity of human bones.

At the east end of the south aisle of Grantham church, Lincolnshire, is what appears to be a charnel vault or crypt: this is of the fourteenth century, and consists of two bays of quadripartite vaulting, groined from a central pier. It is entered from the chancel through a shrine-like structure of rich Perpendicular or fifteenth century work. At the east end of this crypt or vault is an ancient stone altar, indicative of this vault having been a *Capella Carnaria*; an opening in the front was probably for the reception of the sepulchrum with relics. The altar slab is 6 feet $4\frac{1}{2}$ inches long, 3 feet $1\frac{1}{2}$ inches wide, and $4\frac{1}{2}$ inches thick; the upper half is slightly moulded, the lower half has the hollow chamfer. This altar is 3 feet 2 inches in height.

Beneath the fifteenth century vestry on the north side of the chancel of Marldon church, Devon, appears to be a small charnel vault.

At the east end of S. Peter's church, Sandwich, Kent, is a crypt which has evidently been a charnel vault.

At the east end of the chancel of Edmundthorpe church, Leicestershire, is a quatrefoil opening; a similar opening appears in the south wall of the chancel.

Beneath the south transeptal chapel, Norborough church, Northamptonshire, is a singular charnel vault, the entrance to which is by a flight of winding steps from the south-west corner of the chapel; a narrow passage, lighted by a single sloping grated opening, runs along the south side; the vault itself is under the eastern side of the chapels; it is lighted by two sloping and grated openings in the east wall, through which bones might be thrown down. When I saw this charnel vault, in the year 1855, the floor was covered with human sculls and bones. This structure is of the fourteenth century.

Under the eastern portion of the south aisle or chancel of Pakefield church, Suffolk, is a vault, apparently a charnel house, lighted in the east wall by two small and narrow lights, still open, through which bones might be thrown down; in the south wall are indications of one light, now blocked up.

At the east end of the chancel of S. Margaret's church, Lowestoft, Suffolk, is a groined and vaulted crypt of the fourteenth century, formerly apparently a charnel vault, though now used as a vestry. In the east wall of this vault are two small lights. Two piers support the vaulting of this crypt, composed of simple cross quadripartite vaulting in six divisions. The width of this crypt from east to west is 11 feet; the length corresponds with the width of the chancel.

On the north side of the chancel or choir of the parish church of Stratford-upon-Avon, and annexed to it was formerly a crypt or charnel vault, the receptacle of bones, with a *Capella Carnaria* above. Of this I will give the descriptive account of the late Mr. Wheler, the well-known historian of that English shrine, to which the literary pilgrims of every nation are wont to roam. "The crypt or charnel house, though not standing at this time, ought not to be passed over in silence. It was a plain building, 30 feet long and 15 wide, nearly the height of the chancel, and had every appearance of being the most ancient part of the whole church. In this charnel house was contained a vast collection of human bones; how long they had been deposited there is not easily to be determined, but it is evident from the immense quantity contained in the vault it could have been used for no other purpose for many ages. It is supposed the custom was discontinued at the Reformation, as no addition to them had been made in the memory of the oldest inhabitant of the town living in the last century. This vault was built in the unornamented *Saxon Gothic* style (?); the pillars a little above the surface of the earth were each divided into three ribs, intersecting each other, and closed up with unhewn stone. Above was a room, (probably the *Capella Carnaria*), the ascent to which was by a flight of stone steps. In consequence of the dilapidated state of this building, a faculty, at the request of the churchwardens, was granted by the Bishop of Worcester, to raze it to the ground; accordingly the bones were carefully covered over, and the charnel house taken down in the year 1800."

But the most notable, best known, and largest collection of bones in charnel vaults, are those beneath the churches of Hythe in Kent, and Rothwell in Northamptonshire; and these have severally been the origin of most groundless assumptions and unfounded speculations. And first, of the charnel vault at Hythe.

This is a church I have never visited, and can, therefore, only speak of from other sources. In the crypt, said to be a structure of the thirteenth century, beneath the choir of Hythe church, vast quantities of human sculls and bones are deposited, the pile of them being 28 feet in length, and 8 feet in height and breadth. Of these, Hasted, in his history of Kent, gives the following fanciful accounts:—"They are, by the most probable conjectures, supposed to have been the remains of the Britons slain in a bloody battle, fought on the shore between this place and Folkestone with the retreating Saxons, A.D. 456, and to have attained their whiteness by lying for some length of time exposed on the sea shore. Several of the sculls have deep cuts in them, as if made by some heavy weapon, most likely of the Saxons." This conjectural account does not, however, appear to have been in existence in Leland's time, who gives a description of this church in his *Itinerary*, but merely observes of this crypt as "Under the quire a very fayr vaute."

The same historian, Hasted, notices a vault or crypt beneath the neighbouring church of Folkestone, containing a similar collection of sculls and human bones to those deposited in the crypt at Hythe, and he rushes at the inference that from the quantity of them they could not but be from some battle. Those at Hythe he conjectured to be those of the Britons; those at Folkestone to be those of the Saxons. He had evidently never heard of charnel vaults.

The collection of human bones in a crypt beneath the south aisle of Rothwell church, Northamptonshire, has given rise to many imaginary surmises. In the early part of the last century this crypt was unknown, the discovery being made by some workmen employed in making a grave, who broke through the crown of the vault. This crypt is a structure of the fourteenth century, the entrance being at the west end, and a narrow winding passage, with a descent of seventeen steps, led from the porch down into it; the doorway at the west end, though plain, is somewhat singular; the head consists of a horizontal lintel, with a return downwards, and then aslant to the jambs, increasing the width. This crypt is about 36 feet in length and 12 feet in width, with bays of quadripartite vaulting, formed by the intersection of pentagonal-shaped ribs, piled up or ranged at the east end, and on either side, extending to the west end, is a collection of human sculls and bones to the height of upwards of 4 feet, and of the same width. The existence of this crypt or charnel vault for the reception of the bones dug up from time to time in the surrounding cemetery does not appear to have been known to Bridges, the historian; at least, in his printed collections, edited by the Rev. Peter Whalley, nearly at the close of the last century, no notice whatever appears of it.

There are doubtless many more charnel vaults in churches besides those I have enumerated, but the latter will suffice to establish my remarks.

Representation of a Corpse attired for Burial without a Coffin, from a Mural Painting on the wall of the Chapel of the Holy Trinity, Stratford-on-Avon. Late 15th Century.

MURAL PAINTINGS. Before and during the Middle Ages, when the art of printing was unknown, when learning was in a great measure confined to those who led a monastic life and had leisure for study, the knowledge of sacred subjects was conveyed, in no slight degree, to the untutored eye and mind, either by sculptures in bas-relief, or by the paintings which covered, more or less, the walls of our ancient churches. These were the *Biblia Pauperum*, illustrative in the earlier ages of Scriptural subjects, intermixed at a later period with stories originating from legendary lore, from the *legenda aurea*, and other like sources. The system thus practised of the internal decoration of our ancient churches by means of mural paintings, mostly in distemper, is one of considerable antiquity, and may be referred back in our Anglo-Saxon churches to the latter part of the seventh century. Beda, a contemporaneous historian, in his lives of the early Abbots of Monkswearmouth, a monastery founded by Benedict Biscopius, circa A.D. 672, in his account of that, the first, Abbot, informs us that on his return, on one of his journeys, from Rome, "he carried home with him paintings of holy subjects for the ornament of the church of the blessed Peter the Apostle, which he had built at Monkswearmouth: a representation, namely, of the blessed mother of God, and ever Virgin Mary, as well as of the twelve Apostles, which girt the middle '*testudo*' of the same church, a boarding having been run from wall to wall: the figures of the Gospel history with which to decorate the southern part of the church: the images of the visions of the Apocalypse of the blessed John, with which, in like manner, he purposed to decorate the wall on the north:— to the intent that all who entered the church, even if ignorant of letters, might be able to contemplate in what direction soever they looked, the ever-gracious countenance of Christ and His Saints, even though it were in a representation; or with a more wakeful mind might be reminded of the grace of our Lord's incarnation; or, having as it were the strictness of the Last Judgment before their eyes, should thereby be cautioned to examine themselves with the more narrow scrutiny."[s]

Richard, Prior of Hexham, Northumberland, who died A.D. 1190, treating of the original church of Hexham, constructed A.D. 674, by S. Wilfrid, Archbishop of York,

[s] Beda. *Lives of the Abbots Benedict*, &c. Stevenson's Translation.

who employed thereon workmen brought by him from Rome, tells us that the founder decorated this church with histories and images in relief, and with pictures in a variety of colours and with great skill.[h]

By the Synod of Calcuith, held A.D. 816, a representation of the Saint to whom a church or altar was dedicated, was required to be painted either on the wall of the church or on a tablet suspended in the church.

*Seu etiam precipimus unicuique episcopo, ut habeat depictum in pariete oratorii, aut in tabula, vel etiam in altaribus, quibus sanctis sint utraque dedicata.*[i]

Of these paintings, however, and of other early paintings, if any, on the walls of our Anglo-Saxon churches, we have no existing vestiges. The general belief, at the close of the tenth century, that at the expiration of that time the world would come to an end, led to the cessation for a while of church building. We may, however, from examples remaining, trace the progress of mural painting in our churches from perhaps the middle of the twelfth century, or a little subsequent to that period, though the specimens left of that age are but few. Of these early mural paintings, those on the groined roof of the apse in the chapel of S. Gabriel, Canterbury Cathedral, are perhaps the most interesting. Here are depicted our Lord sitting in majesty, within a vesica. On the north, S. Gabriel announcing to Zacharias the approaching birth of the Baptist. On

---

[h] Ipsos etiam et capitella columpnarum quibus sustentantur, et circum sanctuarii historiis et imaginibus et variis celaturarum figuris ex lapide prominentibus, et picturarum, et colorum grata varietate mirabilique decore decoravit.—*Ricardus Prior Hagustaldensis.—Scriptores X. Vol. i. p.* 290.

[i] Wilkins' *Concilia, Vol. i. p.* 169.

the south, S. Gabriel appearing to communicate to S. Mary the incarnation of Jesus. There are also depicted scenes relating to the birth and naming of the Baptist. Seraphim on winged wheels are also represented. Some of the subjects, especially that of the naming of John by Zacharias, are very perfect. The whole composition and architectural details refer' us to the age of the latter part of the twelfth century; it may, however, be doubted whether the painting was by a native artist.[k] At the east end of the church of Copford, in Essex, are paintings of the twelfth century, which have, however, been unfortunately restored : representations of the Apostles and signs of the Zodiac are here given. On the walls and roof of Kempley church, Gloucestershire, within semicircular-headed compartments, are represented the Apostles; these also are of the twelfth century, and well worthy of close examination.

On several of the Norman piers of the nave of the now Cathedral of St. Albans, on the west side, are mural paintings, in fresco or distemper, of the Crucifixion; it is probable that beneath these were altars. These paintings may, I think, be fairly assigned to Walter of Colchester, a famed sculptor and painter, and one of the inmates of the Abbey of St. Albans, whose name has been handed down to posterity by Matthew Paris.

In Charlwood church, Surrey, the legend of S. Margaret is represented in pictorial guise on the wall. These illustrations appear to be of the thirteenth century.

[k] These paintings have been well illustrated in Vol. XIII. of the *Archæologia Cantiana* by the Rev. Canon Scott Robertson.

Many mural paintings were executed in the reign of Henry III., A.D. 1216—1272, who appears to have been a great patron of the fine arts. In 1240, the keepers of the king's works were required to have painted in the church of S. Peter, in the Tower of London, S. Peter in his pontificals and S. Christopher carrying Jesus, an early representation of the legend; and to make two beautiful pictures with the stories of S. Nicholas and S. Catherine, for their respective altars. And in 1242, they were ordered to have the Old and New Testaments painted in the King's chapel at Windsor.

In Battle church, Sussex, in a series of designs in square compartments, in colours red and yellow, is depicted the Passion of Christ. These paintings appear to be of the fourteenth century.

In S. John's church, Winchester, are represented on the walls in fresco or distemper, our Lord within a roundel, sitting in judgment: in four smaller roundels appear the Evangelistic symbols. On the walls of the same church appear the Crucifixion, S. Andrew on the Cross, the Blessed Virgin and infant Christ, and S. Francis. These paintings appear to be of the latter part of the thirteenth or early in the fourteenth century.

The legend of S. Christopher with the infant Christ on his shoulders appears to have been a favourite illustration on the north wall of the nave, especially in the fifteenth century. There are some few examples in the fourteenth century, as in S. John's church, Winchester; Shorwell church, Isle of Wight; and the old church of Croydon, Surrey. Representations of this legend also appear on the walls of East Meon church,

Hants; Melcombe Horsey church, Dorset; of the fifteenth century. S. James, Elmham church, Suffolk; Gawsworth church, Cheshire; Dilleridge church, Wilts; Horley church, Oxon—this is of the fifteenth century; and on the walls of many other churches; the popular belief being that whosoever should behold the figure of S. Christopher would be exempt that day from evil.

"Xp' oferi sancti speciem quicunque tuetur
Illo nempe die nullo langore gravetur."

Issuing from the mouth of one of these gigantic effigies is a scroll inscribed as follows:—

"What art thou that art so hea. . . . . bar I never so hevy a thynge."

From the lips of the infant Christ in reply:—

"I be hevy no wunder nys, for I am the kynge of blys."

Mural paintings of S. Christopher more or less visible are apparent in Burford church, Oxon; and in Morborne church, and Orton Longueville church, Huntingdonshire.

The legend of S. Christopher—an allegory it is supposed of the Schoolmen—became in the fifteenth century a supposed reality, hence we occasionally meet with the precatory words on tombs, *Sancte Christophere ora pro nobis.* By will, dated A.D. 1506, Robert Wolthaite, Vicar of Conisborough, Yorkshire, bequeathed "to the makyng of the church porch xiij*s.* iiij*d*. I wyl that myn executors mak cost of the payntyng of Sent Cristofer in the church of Connysburgh." On the east wall of the north transept of the Cathedral of St. Alban is represented the incredulity of S. Thomas. This is of the fifteenth century. In Rochester Cathedral the wheel of fortune *Rota fortunæ* was represented, an early work of the thirteenth century. S. Michael weighing souls in the balance, sometimes appears, as in Melcombe Horsey church, Dorset, and Dotteridge church, Wilts.

In Stotfold church, Bedfordshire, S. Michael is represented weighing souls, the Blessed Virgin turning the scales. In Preston church, Sussex, S. Michael is also represented weighing souls, and the same subject occurs in Horley church, Oxon. S. George and the Dragon, according to the legend, appears on the walls of Gawsworth church, Cheshire, and S. Gregory's church, Norwich; also in Dartford church, Kent, and on the north aisle of Hornton church, Oxon. Medieval moralities are sometimes represented on church walls, as that of " *Le dit des trois mort, et de trois vifs*," in Ditchingham church, Norfolk, where three kings are represented in their robes, and also as skeletons. The Doom or Last Judgment was sometimes painted over the chancel arch facing west, and sometimes on the west wall of the nave. In the contract for the erection of the Lady chapel, S. Mary's church, Warwick, A.D. 1454, is a covenant "to paint fine and curiously, to make on the west wall the dome of our Lord God Jesus, and all manner of devises and imagery thereto belonging." The west front of the wall over the chancel arch, Trinity chapel, Stratford-upon-Avon, was some years back found to be thus covered; but this painting with others in the same chapel was afterwards obliterated. A painting in fresco or distemper of the Last Judgment, was discovered some years ago on the west face of the wall over the chancel arch, Trinity church, Coventry. To the credit of those then in authority this has been carefully preserved, and the coats of whitewash which tended to conceal it, probably ever since the Reformation, has been judiciously removed.

The murder of Archbishop Becket was also a favourite subject. An early pictorial representation of this event, of the thirteenth century, was some years ago discovered on one of the walls of Preston church, Sussex; and a painting of the same subject on panel, executed in the middle of the fifteenth century, was formerly suspended over or near to the tomb of Henry IV. in Canterbury Cathedral. At the west end of the nave of Trinity chapel, Stratford-upon-Avon, one of the mural paintings in that chapel was this subject, rude in art, of the close of the fifteenth or early in the sixteenth century.

Archbishop Becket is sometimes depicted as a single figure, in his pontificals; he is also so represented in painted glass.[1]

In the year 1804 a series of mural paintings, scriptural, historical, allegorical, and legendary, covering the wall of the chapel of the Holy Trinity, belonging to the Guild of the Holy Cross, Stratford-upon-Avon, were brought to light. Amongst these was depicted the conference of the Queen of Sheba with Solomon. The Last Judgment covered the west face of the wall, divid-

---

[1] By an injunction set forth by royal authority, A.D. 1539, it was ordered,—"That from henceforth the said Thomas Becket shall not be esteemed, named, reputed and called a saint, but Bishop Becket; and that his images and pictures thorow the whole realme shall be pluckt downe and avoided out of all churches, chapels, and other places."—Fox's *Martyrology*.

ing the chancel from the nave. The murder of Archbishop Becket was likewise represented, as also the legendary story of S. George and the Dragon. The invention of the Cross by S. Helena, and subsequent assumed incidents relative thereto, formed a series. These paintings were evidently of a late period; certainly not earlier than the reign of Henry VII., as was evident from the costume of many of the figures, and were of no artistic merit. Before these paintings were obliterated, according to the then prevailing usage, Mr. Thomas Fisher, an antiquary of some eminence, made accurate coloured drawings of them, which on a scale of an inch and a half to a foot, appear to have been first published in 1808, and again in a folio volume in 1836. The colours used in these paintings were diverse, red, yellow, blue, black, green, and white; but mural paintings of an earlier period, *viz.* of the fourteenth century, were often monochromatic only, outlines in red being not unusual.

Mural paintings appear also on the face of walls at the back of tombs when placed beneath an arch in the wall, as at the back of the tomb in Ingham church, Norfolk, of Sir Oliver Ingham, who died A.D. 1344; and at the back of a tomb in Dodford church, Northamptonshire, where angels are pourtrayed carrying a soul to heaven in a winding sheet. This was a conventual mode of representation from the twelfth to the early part of the sixteenth century, and it is thus alluded to by Becon, one of our early Reformers, in his treatise *The Acts of Christ and of Antichrist*, A.D. 1564,—" Christ was buried in a poor monument, sepulchre, or grave, without any funeral pomp." "Antichrist is buried

Painting at the back of a Tomb in Dodford Church, Northamptonshire. 14th Century.

in a glorious tomb, well gilt and very gorgeously set out with many torches and with great solemnity, and with angels gloriously portured that bear his soul to heaven."

PAINTINGS ON ROOD-LOFT SCREENS AND PARCLOSES. Besides the mural paintings, which were common, the rood-loft screens and parcloses in many of our churches appeared to have been covered with painting and gilding. Of these the churches in Norfolk and Suffolk present the most perfect examples we now possess. The lower portions of these screens being panelled, the various compartments were painted, apparently in *tempera*, with single figures, mostly of saints, or reputed saints, of the medieval church, distinguished by their several symbols. Sometimes the Hierarchy of Angels was thus depicted.

S. AGATHA V. M.[m] (A.D. 251), with a knife at her breast, Wiggenhall church, Norfolk.

S. AGNES, V. M. (A.D. 304), with a sword and lamb, on the rood-screen at North Elmham, N.[n]; Westhall, S.[o]; and Eye, S.

S. ANDREW, Apostle with the Cross Saltire, ✕, is represented on the rood-screens of Worstead, N.; Ranworth, N.; Lessingham, N.; Tunstead, N.; Edingthorpe, N.; and Blofield, N.

S. ANN teaching the Blessed Virgin to read, on the rood-loft of Houghton le Dale,

S. ANTHONY, (A.D. 250), with staff, pig, and bell, on the rood-loft at Westhall, S.

S. APOLLONIA, V. M. (A.D. 250) holding a tooth in

[m] V. M. Virgin and Martyr.   [n] N. Norfolk.   [o] S. Suffolk.

pincers, on the rood-screens of Ludham, N.; Westhall, S.; Barton Turf, N.; and Lessingham, N.

S. BARBARA, V. M. (A.D. 306), carrying a tower, on the rood-screens of North Walsham, N.; Barton Turf, N.; Filby, N.; and Yaxley, S.

S. BARTHOLOMEW, Apostle, with a flaying knife in his hand, on the rood-screens of Edingthorpe, N.; Tunstead, N.; Ranworth, N.; and Worstead, N.

S. BENEDICT, Abbot, (A.D. 543), with a demon on each side of him, piercing one of them with his pastoral staff, on the rood-screen, Burlingham, S. Andrew, N.

S. BLAISE, B. M. (A.D. 304), with mitre and crosier, on the rood-screen at Hempstead, N.

S. BRIDGET of Sweden, wid. (A.D. 1373), crowned, crosier, book and chain in her hand, on the rood-screen, Westhall, S.

S. CATHERINE, V. M. (A.D. 290), with wheel set with spikes, or sword, on the rood-screens at North Walsham, N.; Westhall, S.; Lessingham, N.; and Filby, N.

S. CECILY, V. M. (A.D. 220), crown, wreath of flowers, and a palm, rood-screens Filby, and Burlingham, St. Andrew, N.

S. CLEMENT, P. M. (A.D. 100), mitre, triple cross, and anchor, rood-screen, Westhall, S.

S. DIONYSIUS or DENIS, B. M. (A.D. 272) carrying his head in his hand, rood-screen, Grafton Regis, Northamptonshire.

S. DOROTHY, V. M. (3rd century), fruit and flowers, rood-screens, North Elmham, N.; Blofield, N.; Yaxley, S.; Westhall, S.; Trimmingham, N.; and Walpole, St. Peter, N.

S. EDMUND, K. M. (A.D. 870), pierced with arrows, rood-screen, North Walsham, N.

S. EDWARD, K. Confessor (A.D. 1066), sceptre and ring rood-screens, Ludham, N.; Stalham, N.; Barton Turf, N.; Attleborough, N.; and Eye, S.

S. ERASMUS, B. M. (A.D. 303), windlass in his hand, bowel wound round it, rood-screen, Hempstead, N.

S. ETHELDREDA, V. (A.D. 679), crowned, with crosier, rood-screens, Burlingham S. Andrew, N.; Ranworth, N.; Upton, N.; Oxburgh, N.

S. FRANCIS of Assis, C. (A.D. 1226), crowned with thorns, stigmata on his hands, feet, and side, rood-screens, Stalham, N., and Hempstead, N.

S. GILES, Abbot, (7th century), hind lying at his feet, rood-screen, Lessingham, N.

S. GREGORY the Great, P. C. D. (A.D. 604), double-barred cross, &c., rood-screens, Ludham; Tunstead, N.; East Ruston, N.; Houghton le Dale; Lessingham, N.

S. HELEN, Empress, (A.D. 328), bearing a cross, rood-screen, Eye, S.; south parclose, Ranworth, N.

S. JAMES the Greater, Apostle, pilgrim, with staff, shell, hat, and wallet, rood-screens, Tunstead, N.; Lessingham, N.; Worstead, N.; Edingthorpe, N.; Blofield, N.; Ringland, N.; Ranworth, N.

S. JAMES the Less, Apostle, a fuller's club in his hand, rood-screens, Belaugh, N.; Ranworth, N.; Lessingham, N.; Blofield, N.; Worstead, N.; Ringland, N.; and Tunstead, N.

S. JEROM, C. D. (A.D. 420), cardinal's hat, or robes, or both, rood-screens, Lessingham, N., and East Ruston, N.

S. JOHN the Baptist, lamb on a book, &c., rood-screen,

210  INTERNAL ARRANGEMENT OF CHURCHES

Burlingham, S. Andrew, N.; parclose screens, Ranworth, N.; Worstead, N.

S. JOHN, Apostle and Evangelist, chalice, with serpent issuing from it, rood-screens, Worstead, N., and Ranworth, N.

S. JUDE, Apostle, a boat in his hand, rood-screens, Ringland, N.; Lessingham, N.; Belaugh, N.; Worstead, N.; Swafield, N.; and Tunstead, N.

S. LAWRENCE, M. (A.D. 258), deacon, holding a gridiron, rood-screens, Ludham, N., and Hempstead, N.; screens, Worstead, N., and Ranworth, N.

S. MARGARET, V. M. (4th century), piercing a dragon, rood-screens, North Walsham, N.; Filby, N.; Lessingham, N.; and Westhall, N.

S. MARY MAGDALEN, with a box of ointment in her hand, on the rood-screens of Oxborough, N.; Leasingham, N.; Ludham, N.; North Walsham, N.; Yaxley, N.; and Bramfield, S.

S. MATTHIAS, Apostle, bearing a halbert, rood-screens, Aylsham, N.; Blofield, N.; Ringland, N.; Tunstead, N.

S. MATTHEW, Apostle and Evangelist, holding a money bag, or money box, rood-screens, North Walsham, N.; Worstead, N.; Ringland, N.; Aylsham, N.

S. MICHAEL, Archangel, in armour combating a dragon with a sword, Ranworth, N., parclose, with scales weighing souls, rood-screen, Filby, N.

S. PAUL, Apostle, resting upon or holding a sword, rood-screens, Aylsham, N.; Lessingham, N.; Belaugh, N.; Tunstead, N.; Ranworth, N.; Filby, N.; Edingthorpe, N.

S. PETER, Apostle, single or two keys, rood-screens,

Edingthorpe, N.; Tunstead, N.; Lessingham, N.; Irstead, N.; Filby, N.; Worstead, N.; Ringland, N.; Ranworth, N.; Swafield, N.; Westwick, N.

S. PETRONILLA, V. (1st century), key and clasped book, rood-screens, North Elmham, N.; Trimmingham, N.

S. PHILIP, Apostle, basket, with bread, rood-screens, Marsham, N.; North Walsham, N.; Ringland, N.; Irstead, N.; Lessingham, N.; Tunstead, N.; Belaugh, N.; Worstead, N.; Blofield, N.

S. ROCH, C. (A.D. 1327—48), Pilgrim, plague spot on thigh, rood-screen, Stalham, N.

S. SIMON, Apostle, with a fish or two in his hand, rood-screens, Ranworth, N.; Blofield, N.; Worstead, N.; North Walsham, N.; Aylsham, N.; Swafield, N.; Belaugh, N.; Tunstead, N.; with an oar in hand, Southwold, S.; Lessingham, N.; with a fuller's bat, Ringland, N.; Causton, N.; with a saw in his hand, rood-screen, Sotterley, S.

S. SITHA, V. (circa 870), rosary, bag, book, keys, or some or one of them, rood-screens, North Elmham, N.; Barton Turf, N.; Westhall, S.

S. STEPHEN, Protomartyr, as deacon holding stones, north parclose, Ranworth, N.; rood-screens, Ludham, N.; Hempstead, N.

S. THOMAS, Apostle, with a spear or lance, rood-screens, Ranworth, N.; Blofield, N.; Swafield, N.; Belaugh, N.; Ringland, N.; Tunstead, N.; Sotterley, S.

S. THOMAS of Canterbury, (A.D. 1170), pall and crosier, rood-screens, Sparham, N.; Stalham, N.; Burlingham S. Andrew, N.; screens, Attleborough, N.; Worstead, N.

S. WALSTAN, (A.D. 1016), crown, sceptre, and scythe,

one or more; rood-screens, Burlingham S. Andrew, N.; Ludham, N.; Sparham, N.; Barnham Broom, N.

Such were the saints or reputed saints more or less held in repute in the churches of this country prior to the middle of the sixteenth century, and the representations of whom, with their various emblems, were thus painted on panel-work in the several churches thus adduced.

Besides these, ANGELS are depicted on the rood-screens of Barton Turf, N., and Southwold, S.

ARCHANGELS, at Barton Turf and Southwold.

CHERUBIMS, clothed with feathers, and hands uplifted in adoration, on the rood-screen at Barton Turf; and, standing on a wheel, with hands folded on the breast, on the rood-screen at Southwold.

SERAPHIMS, with a thurible, and clothed with feathers, on the rood-screen at Barton Turf; and, standing on a wheel, with a scroll in the left hand, inscribed, *Scūs, Scūs, Scūs*, on the rood-screen at Southwold.

THRONES are represented on the rood-screen, Barton Turf, as a throne and golden scales; and on the rood-screen at Southwold, as a tower in hands, and belt of squares, all of gold.

DOMINATIONS at Barton Turf, with a triple crown and chasuble; and at Southwold, with the chalice and host in right hand, and globe and cross in left hand.

PRINCIPALITIES are represented at Barton Turf as crowned, bearing a palm branch in the left hand, and a vial in the right; and at Southwold, as standing in a citadel with a sceptre.

POWERS are represented both at Barton Turf and Southwold as chaining and scourging devils.

VIRTUES are represented at Barton Turf in green clothing, with blue feathers; and at Southwold, with a crown in the right hand, and censer in the left hand.*

I have not met with any parclose, or rood-screen paintings, of a period anterior to the fifteenth century.

There were other paintings on board or panel of historic or legendary scenes with which our churches were formerly decorated. Of these very few are now remaining. These, in old inventories, were called "painted tables." At the back of the stalls in Carlisle Cathedral is a painting of this description, in which is depicted incidents from the life of S. Anthony the hermit, and others from the life of S. Cuthbert. There are also figures of the twelve Apostles, each with the sentence from the Creed above him. On this painting are the initials of Thomas Gondibour, who became Prior of Carlisle in 1484, to about which period, from the costume and other accessories, this painting appears to have been executed.

In the Conventual church of Romsey, Hants, is a curious but mutilated specimen of a "painted table" of wood or panel of the latter part of the fifteenth century, which appears to have been at the back of some altar in that church.

A painted table of wood, representing the murder, A.D. 1170, of Thomas Becket, Archbishop of Canterbury, formerly hung against the columns at the head of the tomb of Henry IV., in Canterbury Cathedral. The

---

* The above description of saints, with their emblems, and of the Heavenly Hierarchy, has been derived from that valuable work, *Emblems of Saints*, by F. C. Husenbeth, D.D., V.C.

archbishop is represented in his cassock and mantle, or cope; the four knights are depicted in vizored helmets, jupons and armour of the latter part of the reign of Edward III., or that of Richard II. It is the earliest instance of an historic painting on wood, being of the latter part of the fourteenth century, I have hitherto met with in our churches. Whether this "painted table" is still in existence, I know not.[q]

PAINTED GLASS. Without tracing the origin of flat or plate glass with which windows were anciently glazed, that kind of glass appears to have been known in this country by the Romans, since fragments of it have been found on the site of Roman stations in juxtaposition with other Roman remains.[r] Vessels of Roman glass are not uncommon.

Glass windows may be traced to the fifth century.

In the latter part of the seventh century, when Archbishop Wilfred was employed about the cathedral at York, and Benedict Biscopius was building the Conventual church at Monkswearmouth, both are recorded to have sent to France for artists to glaze the windows of those churches.

[q] At the foot of the coloured engraving is the following account:—"Engrav'd by T. Carter from his restored drawing of the defac'd parts of the original painting on board hung against columns at the head of the tomb of Henry IV. in Canterbury Cathedral, copied by him in its present state in a former drawing, both of which are now in the possession of R[d]. Bell, Esq[r].
"Pub[d] as the act directs by J. Carter, Wood S[t]., West[r]., July 1, 1786."

[r] I have in my possession a fragment of flat or plate glass, 4 inches in length, of a medium width of 1¾ of an inch, of a greenish hue, and of the same degree of opaqueness as ground glass, but with a certain degree of lucidity. This was found on the site of the ancient Roman station Tripontium, at Cave's Inn, adjoining the Watling street Road, in the parish of Churchover, in the County of Warwick.

It was not however till the very close of the twelfth, or early in the thirteenth century, that we can trace the introduction of coloured glass in pattern mosaic work, or pictorial subjects in the windows of our larger churches, as in those of the cathedrals of Canterbury and York, in which still exist the most ancient existing remains of coloured glass we have in this country. Anciently the glass was coloured throughout by oxides of metal fused with it; hence the term "pot metal" glass; and pieces of glass thus prepared were formed into geometrical patterns. In York Cathedral, where we have perhaps the most ancient coloured glass in compartments forming mosaic work, the colours consist of a greenish white, green, blue, yellow, and pink, of pot metal glass, and flashed ruby. One of the earliest pictorial representations on painted glass where figures are grouped, is in the upper half of the large north transept window, Canterbury Cathedral. This is divided into nine compartments, three of which are medallions, four square and two half medallions. In these are depicted the following scenes, both from the Old and New Testaments. Joseph as Governor of the land of Egypt, with his Brethren. The Exodus of the Israelites. Balaam on his Ass. The Queen of Sheba and King Solomon. The Magi on horseback. The Magi before Herod. Offering of the Magi.

The predominating tone of the glass is a deep azure, occupying the background of the different groups; the other colours are yellow, green, red, white, and puce. Bordering some of the groups are inscriptions. This glass may, I think, be referable to the early part of the thirteenth century. But the fragment of pictorial glass in the same cathedral, (of which an illustration is here given), represents two knights in mail armour, without surcoats, with nasal helmets and kite-shaped shields, in the military costume of the twelfth century, to the close of which this fragment may perhaps be assigned. In Dorchester church, Oxfordshire, is a fragment of early pictorial glass of the thirteenth century, representing an archbishop, S. Augustine; bishop, S. Birinus; the first bishop of that ancient see, and a third figure. The archbishop wears the pall and holds a crozier, the bishop holds a pastoral staff. Another design in the same church represents a priest and deacon, the former with the host, the latter bearing the ampullæ. In the designs for mosaic or pattern glazing we may trace the age of its greater prevalence in the thirteenth century by its stiff-leaved foliage, which may be compared with the sculptured foliage of the same period. The windows of the five lights in the north transept of York Cathedral are full of stiff foliage formed of white glass, the other colours forming the geometrical designs are red, green, yellow, and blue. Of the thirteenth century is some ancient stained glass in Chetwode church, Bucks. Here we have mosaic medallions with stiff white foliage, whilst in pointed oval compartments, forming the well-known symbol *vesica piscis*, are single figures of saints and crowned personages, each clad in a vest and mantle, colours yellow and green, with blue or azure backgrounds.'

' The mosaic pattern window lights in York Cathedral are engraved in Brown's *History of the Metropolitan Church of S. Peter, York*. The partly mosaic and partly figure window lights at Chetwode church are engraved in Lyson's *Magna Britannia*, Vol. i. p. 488.

Fragment of Painted Glass, from a window in the north aisle of the choir of Canterbury Cathedral, shewing armour of the twelfth century.

PREVIOUS TO THE REFORMATION. 219

The mosaic windows of the fourteenth century are hardly so numerous as those of the preceding age, and in the white glass, instead of the stiff-leaved foliage, we find the oak leaf and acorn pattern. Mosaic and pattern glass of this period may be found in York Cathedral, in Chesham Bois church, Bucks, and Norbury church, Derbyshire. The east window of Long Itchington church, Warwickshire, was formerly filled with pattern glazing of this century, but within the last few years this has disappeared.

Pattern Glazing, Long Itchington Church, Warwickshire.

In the fourteenth century, single figures beneath pedimental canopies, of chaste design, are frequent. In the windows of Tewkesbury Abbey church, Gloucestershire, are several single figures of this description, some of knights in armour. The stained glass in the windows of Merton College chapel, Oxford, exhibits single figures under pedimental canopies: over the head of one of these, the kneeling figure of one habited in a cowl, is a scroll inscribed "*Magister Henricus de Mammersfeld me fecit.*" In the chancel of Stanford church, Northamptonshire, are single figures of the Apostles in painted glass, each appearing beneath an ogee-headed canopy, cinquefoiled within the head and crocketted externally, and the sides of the canopy are flanked by pinnacled buttressets in stages. The east window of Norbury church, Derbyshire, contains figures of the twelve Apostles, each with a scroll containing a portion of the Apostles' Creed: the side windows are filled with pattern glazing, in which portions of stained glass and armorial bearings are introduced. Of this century we sometimes come across a Jesse tree or window, so called. Portions of one of these removed from Merivale Abbey church, Warwickshire, are now preserved in the chapel of the gatehouse to that Abbey, and in the neighbouring church of Mancetter. An early specimen of the Jesse window is in York Cathedral. The remains of a Jesse window is in Westwell church, Kent; and there is a fine Jesse window of this era in Lowick church, Northamptonshire.

The predominating tone of colour of painted or stained glass of the fourteenth century was ruby. Throughout this century it was a very prevalent custom to decorate church windows, especially within the flowing tracery of the upper lights, with plain heater-shaped shields of glass, heraldically charged with the armorial bearings of many families connected with the parish or church. These charges were sometimes simple, sometimes parted per pale, and sometimes quartered. When Sir William Dugdale, in the middle of the seventeenth century, published his *Antiquities of Warwickshire*, these heraldically charged shields in the windows of the churches of that county were very numerous: few, however, are now remaining; amongst which are some in the windows of Bilton church in that county. Burton, in his *History of Leicestershire*, also notices many heraldic shields then existing in the windows of the churches of that county.

The toning down of the original lucid colours in stained glass windows of the thirteenth and fourteenth centuries, so as to dim their lustre and occasion a certain degree of opaqueness, has been the action of the weather and rain-beating for centuries on the external surface of the glass, and thereby causing more or less of corrosion.[1]

The painted glass of the fifteenth century exhibits in its general tone a greater degree of lucidity than what is apparent in the coloured glass of the two preceding centuries. This was occasioned by the admixture of a larger quantity of white or plain glass. Diamond-shaped quarries of white glass, with some device in yellow stain in the centre, as also initial letters in the same yellow stain, were common. This was applied to pattern glazing. Of figure or pictorial glazing, portions which represented human flesh were of white glass, with the features de-

[1] I have in my possession fragments of painted glass of the thirteenth and fourteenth centuries, on the external surface of which numerous small holes appear as if eaten into by the action of the weather.

lineated in outline. Besides single figures beneath canopies, more elaborate than in preceding ages, but not so chaste, we sometimes meet with legendary stories of local saints in painted glass, more or less perfect, as in Great Malvern church, Worcestershire, in some of the windows of which is represented the legend of S. Werstan, a local hermit saint of whom little is known, and of his martyrdom." In Morley church, Derbyshire, the legend of " Saynt Robert," a local saint, is depicted in painted glass, said to have been brought from Dale Abbey. In S. Neot's church, Cornwall, the legend of S. Neot, and the legend of S. George.*

The windows of S. Neot's church also retain a quantity of painted glass from 1400 to 1532. These represent various subjects from the Old Testament, as the Creation of the World, the Fall of Adam, the Building of Noah's Ark; and the nine grades of the Angelic Hierarchy.

The Beauchamp chapel, S. Mary's church, Warwick, erected A.D. 1443—1464, being twenty-one years in building, was enriched as to the windows with painted glass. By a contract, entered into the 23rd day of June 25, Henry VI., John Prudde, of Westminster, Glasier, covenanted " to glaze all the windows in the new chappell at Warwick with glasse beyond the seas, and with no glasse of England; and that in the finest wise, with the best, cleanest, and strongest glasse of beyond the sea that

---

" An essay on this glass, with illustrations, from the pen of the late Mr. Albert Way, F.S.A., appears in the second volume of the *Journal of the Archæological Institute.*

* The several subjects of these legends are treated of in Gorham's *History of S. Neots.*

may be had in England, and of the finest colours of blue, yellow, red, purpure, sanguine and violet, and of all other colours that shall be most necessary and best to make rich embellished matters, images and stories, that shall be delivered and appointed by the said Executors by patterns in paper, afterwards to be newly traced and pictured by another painter in rich colour at the charges of the said glazier. All which proportions the said John Prudde must make perfectly to fine glase, eneylin it, and finely and strongly set it in lead and souder as well as any glasse is in England. Of white glasse, green glasse, and black glasse, he shall put in as little as shall be needful for the shewing and setting forth of the matters Images and storyes. And the said Glasier shall take charge of the same glasse, wrought and to be brought to Warwick, and set up there in the windows of the said Chappell, the Executors paying to the said Glasier for every foot of glasse iis., and so for the whole, xci*li*. is. x*d*."

In this contract the designs to be carried out are not specifically noticed, but merely alluded to generally, and we have only fragments now remaining to enable us to form an opinion on the appearance of the glass when originally completed. The foreign glass, then considered superior to the English made glass, was probably pot metal only.

Amongst the fragments of this glass now remaining, we find angels depicted in albs, with apparelled amices, with circlets round their heads, surmounted by a cross in front over the forehead of each. Some appear in the act of waving thuribles, others are represented as playing on different instruments such as the crowde, a stringed instrument with a bow, the precursor of the violin, on pipes, bagpipes, and other wind instruments, and organs. Some are represented with a scroll before them on which musical notes are pricked, Seraphim are feathered over like birds, and Cherubim holding scrolls and standing on wheels, in allusion to the mystical vision of Ezekiel, their countenances being represented of a bright red colour. Amongst the fragments is the head of our Saviour, crowned with thorns and encircled by a nimbus, charged with a cross. S. Thomas of Canterbury is depicted as vested in a white alb, a purple tunic, a crimson coloured dalmatic, and a rich cope, with a mitre on his head, an aumasse or furred hood about his neck and shoulders, and a crozier in his hand. S. John of Bridlington is represented as an abbot, bareheaded, vested in the aumasse and cope, and holding a pastoral staff, which is headed with a rich crook turned inwards. One figure, said to represent S. Alban, the British proto-martyr, is depicted in armour of plate over which the tabard is worn, and on his back is a rich mantle, fastened by a cordon in front: of his armour, the genouilleres, jambs, and sollerets, (which latter are pointed at the toes), are all that is visible; on his head he wears a cap of estate; in his right hand is a staff, whilst in his left he holds a pent-house covered cross, similar in design to many of the small ancient sepulchral crosses formerly placed at the heads of graves.

The church of Fairford, in Gloucestershire, which was commenced building by John Tame, in 1493, and which was finished after his death in 1500 by his son, is celebrated for its painted glass, probably executed at the very close of the fifteenth, or at the commencement of the sixteenth century. Whoever may have designed this glass, respecting which there have been controversies, the windows of the church, twenty-eight in number, are fairly well filled, and the glass well merits a study. Many of the subjects are taken from the Old and New Testaments. Of the former there are but four: 1, the Temptation of Eve; 2, the Lord appearing to Moses in a fiery bush; 3, the sign given to Gideon; 4, the Queen of Sheba's visit to Solomon. The subjects from the New Testament comprise: the Annunciation; the Nativity; the adoration of the Magi; the presentation in the Temple; the flight into Egypt; Christ disputing with the Doctors; Christ's entry into Jerusalem; Christ in the Garden of Olives; Pilate washing his hands; the scourging of Christ; Christ bearing His cross; the Crucifixion; the descent from the Cross; the Entombment; Christ appearing to Mary Magdalen; Christ and His disciples at Emmaus; Christ appearing to His disciples; the incredulity of S. Thomas; the miraculous draught of fishes; the Ascension; and the descent of the Holy Ghost;—subjects enough to illustrate a pictorial New Testament. Besides the above are represented the twelve Apostles; the four Evangelists; the four primitive Fathers of the Church; and the twelve Prophets. The Apocryphal Gospels are also to some extent illustrated. So much glass of a particular period is here displayed that it is well worth making a pilgrimage to Fairford for a careful study of this glass, of a period just preceding the Renaissance.

In the early part of the sixteenth century, and previous to the close of the reign of Henry VIII., the style in sculpture and painting known as that of the Renaissance was gradually making its way into this country; and the glass in the windows of King's College chapel, Cambridge, are, perhaps, the most perfect of this period we possess.

The foundation of this chapel was commenced A.D. 1446; owing to various causes, however, although of Royal foundation, it does not appear to have been finished as to its internal fittings, *viz.*, the screen and part of the stalls, till the year 1534. Previous to this, however, in the 18 Henry VIII., (A.D. 1527), Galyon Hone, Richard Bounde, Thomas Reeve, and James Nicholson, all of or near London, glasiers, covenanted to " glase and set up at their proper costs and charges eighteene wyndowes of the upper story of the great churche within the Kynges College of Cambridge, whereof the wyndowe in the este ende of the same churche to be oon and the wyndowe in the west ende of the same churche to be another, and so seryatly the residue w$^{th}$ good, clene, sure, and perfyt glasse and oryent colours, and imagery of the story of y$^e$ olde lawe and of y$^e$ newe lawe, after y$^e$ fourme, maner, goodnesse, curyousytie, and clenelyness in every poynt of y$^e$ glasse wyndowes of y$^e$ kynge's newe chapel at Westmynster; six of y$^e$ seid wyndowes to be clerely set up and fynyshed after y$^e$ fourme aboveseid within twelve monthes from y$^e$ date of y$^e$ indenture, and y$^e$ twelve wyndowes residue to be clerely set up and fully fynyshed within foure years after next ensuing; and further, y$^e$

said Galyon, Richard, Thomas, and James covenanted well and surely to bynde alle the seid wyndowes with double bandes of leade, for the defense of greate wyndes, and outragious wetherings." The glasiers were to be paid "for the glasse, workmanship, and setting up of every foote of the seid glasse by them to be provided, wrought, and set up after the foùrme aboveseid, sixteue pence sterlinge."[y]

These windows might well form the subjects for a Pictorial Bible, the types being taken from the Old, the anti-types from the New Testaments. Some of the designs remind one of the School of Raphael.[z] They are, perhaps, the latest series of pictorial windows we possess anterior to the Suppression of the Monasteries, and destruction, wholly or partially, of the Conventual churches.

In S. Mary's church, Shrewsbury, many of the windows are filled with ancient painted glass of different periods, collected and placed there by, I believe, a late vicar, the Rev. D. Rowland. The east window contains the principal remains of a Jesse window, removed there from the old demolished church of S. Chad. These windows are well worthy of study.

The east window of Twycross church, Leicestershire, contains some interesting fragments of ancient figured glass of the thirteenth century, removed during the French Revolution from some of the windows of La

[y] Le Keux's *Memorials of Cambridge*, Vol. xi. p. 43.

[z] A copious Essay on these windows, with a full explanation of the subjects they illustrate, by Mr. W. J. Bolton, is given in the 12th Vol. of the *Journal of the Archæological Institute*. Artistic notes on these windows, by Mr. G. Scharf, Junior, with illustrations, also appear in the same Vol.

Sainte Chapelle, Paris; and the east window of Rugby School chapel contains, amongst other glass, a fine representation of the Adoration of the Magi, a German design, apparently of the school of Albert Durer. This, a fine specimen of the Renaissance, came from the church of Oischot, near Louvain.

PAVEMENTS OF ENCAUSTIC TILES. Besides the decorative paintings on church walls, and the windows filled with painted glass, the pavement, or portions of it, in many of our churches was not without a diversity of colour and embellishments, as laid down in encaustic tiles. Of these pavements we have but few, if any, existing in a perfect state. Like as with the scattered remnants of ancient painted glass, we find them mostly in a fragmental state, sometimes broken up to be replaced by sepulchral slabs, sometimes from causes not explainable; and we are, therefore, better able to treat of the tiles, of which they are composed, severally, than in any composite arrangement.

We rarely meet with encaustic tiles used in the pavement of churches distinguished by any well-known characteristic device earlier than the latter part of the thirteenth century. Throughout the fourteenth and fifteenth centuries they prevailed; they became less common in the early half of the sixteenth century, towards the middle of which the laying of them down appears, with the declension of church architecture, to have fallen into disuse.

They are found in floriated patterns, four or more tiles completing the design. Heraldic devices are common, heater-shaped shields, bearing generally simple charges, though sometimes parted per pale, and sometimes quartered; these are mostly of the fourteenth century. We also find the Lion rampant and the Spread Eagle; arms and badge of the King of the Romans; Warriors on horseback; Civilians on foot; Kings' heads; the Alphabet; a Fish in the vesica, symbolical of our Saviour; and numerous other devices, represented for the most part in two colours only, red and yellow.

For the tile being in the main of red clay, was, whilst in a plastic state, impressed with a device, in the sinking, formed by which, yellow clay was inserted; and we rarely find in encaustic tiles more than these two colours, red and yellow. The surface of these tiles was generally, perhaps not always, glazed.

At Woodperry church, Oxfordshire, are portions of a pavement of encaustic tiles of pattern design, of red and yellow in colour. Four tiles form a device consisting of a quatrefoil, within which are four fleur-de-lis arranged crosswise. Outward of the quatrefoil, at the four corners of the design, is the trefoil leaf foliage. Portions of the pavement exhibit birds with expanded wings, rudely delineated. This pavement appears to be of the thirteenth century.[a]

In or about the year 1854, in excavations made on the site of Chertsey Abbey, Surrey, remains of an interesting pavement, composed of encaustic tiles, were discovered. These displayed subjects taken from the Old Testament, such as David slaying the Lion, and David before Saul. Also the representation of a conflict be-

[a] Engraved in Vol. iii. of the *Journal of the Archæological Institute*, p. 129.

tween a Knight and a Lion, and other devices. These tiles were supposed to be of the thirteenth century.

At the east end of the site of the Conventual church of Neath Abbey, Glamorganshire, was discovered, in 1803, on some excavations being made, a rich pavement of encaustic tiles, consisting of heraldic charges on shields; colours red and yellow. Another portion of the pavement represented Stags running, and a Forester on foot blowing a horn. Another portion simply consisted of pattern tiles. This pavement was of the fourteenth century.[b]

In Higham Ferrers church, Northamptonshire, is a tile pavement of original arrangement on the steps which led to the High Altar. These are covered with indented or encaustic tiles laid in various patterns. One of these is a lozenge, formed by a square black tile, scored in squares, as a centre, surrounded by four narrow yellow bordering tiles, having a small black one at each angle. On another portion lozenge-shaped tiles are laid down.

Encaustic tiles, in a variety of devices, may be found in many churches, and on the sites of others now demolished.

Of particular tiles, I may enumerate a few from a small collection, from different sources, in my possession.

One of these is larger than those generally met with, being $6\frac{3}{8}$ inches square and 2 inches in thickness; this has stiff-leaved foliage in relief; it is monotone, of a dark green or lead colour. It was found in the precincts

[b] Engraved in Vol. iii., *Journal of the Archæological Institute*, p. 277.

of Combe Abbey, Warwickshire, and I should assign it to the thirteenth century.

A second, of more common size, $4\frac{5}{8}$ inches square, exhibits a warrior on horseback, with heater-shaped shield and lance; the ground of the tile is red, and the device yellow,—the latter is very rudely delineated. This tile is glazed; it is apparently of the thirteenth century, and was found at Brinklow, in Warwickshire.

A third, $4\frac{7}{8}$ inches square, exhibits on a red ground the figure of a civilian in the costume of the fourteenth century, clad in a close-fitting tunic or cote hardi reaching to the thighs, with the hood worn on the head and over the shoulders; close-fitting pantaloons or hose envelope the thighs and legs, and long liripipes fall from his close-fitting sleeves. In his left hand is held a roundel of some description, whilst a dog and trees figure as accessories. The design is delineated in white or yellow clay, on a red ground. Where this tile was from I have no knowledge, but I have met with similar tiles; it is of the fourteenth century.

A fourth tile, 5 inches square, exhibits the head of a king, placed on the tile lozengewise; the crown is worn on the head, the face is close shaven, and the arms and hands are stretched out upwards; the ground of the tile is red, the device yellow. This tile was found at Combe Abbey or Brinklow, and I should assign it to the latter part of the fifteenth century. It is different to other tiles I have met with of crowned heads, and which are of an earlier period.

A fifth tile, $4\frac{3}{4}$ inches square, of a monotone, presents the profile of a face in incised outline, with a quasi-classic helme on the head, somewhat similar to those heads within roundels we sometimes meet with in wood panel-work of the early part of the sixteenth century, to which period I should ascribe this to be. It came from S. Nicholas church, Brighton.

I have several other encaustic tiles in my possession, heraldic and otherwise, but of no particular interest.

Kilns for the manufacture of this particular description of fictile ware have been discovered at Great Malvern; in the parish of S. Mary, Witton, near Droitwich; and at Great Bedwyn, Wilts.[c]

The several foregoing sectional details, descriptive of the internal arrangement of our parish churches previous to the Reformation, may convey a general though by no means exhaustive idea of their appearance during the Middle Ages. Upon the sepulchral monuments they contained I have not attempted to descant. Of these, from the thirteenth to the middle of the sixteenth century,—for of a later period I shall have subsequently to treat,—our ancient churches present examples of extreme interest. These consist of the plain coffin-shaped slab, without device or inscription, wider at the head than at the foot; of the lowly coped tomb; of the sculptured effigy beneath a mural-arched recess; of the high tomb, more or less elaborately worked on the sides with architectural and heraldic detail, and with or without effigies on the covering slab; and of incised brasses of individuals of different callings in life, more especially of the clergy, in their Eucharistic vestments or choral

[c] *Journal of the British Archæological Association, Vol. iv. p. 216.*

habits, on both of which I shall hereafter take occasion to remark.

Sepulchral monuments are indeed in many of our ancient churches, the only remaining objects of antiquity we can regard with satisfaction. In the so-called restorations of our churches during the last half century—without denying that conservative restoration was needed—many details of historic interest, bearing on points of ancient church governance and discipline, and which ought to have been carefully preserved, have been needlessly and ruthlessly swept away. Monuments have been removed from their original positions, and set apart from the remains of the individuals over which they were erected. Of the commencement of the changes generally in our churches we must have reference to the suppression of the monasteries and religious houses. Of the interesting ruins of these, and of the conventual churches attached to them, differing in some respects in their internal arrangement from mere parish churches, I purpose next to treat.

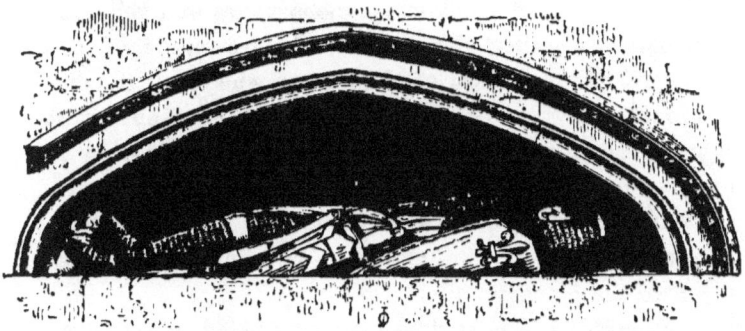

Sepulchral Arch and Effigy, 14th Century, Coleshill Church, Warwickshire.

Gable Cross, Colwall Church, Herefordshire.

## CHAPTER II.

### OF MONASTIC ARRANGEMENT.

TO the untutored eye and mind the numerous existing remains, scattered all over the kingdom, of our ancient monasteries and religious houses,—whether Benedictine, Cistercian, or of the several other Orders of Cœnobites, who dwelt in common, and had each their particular rule,—present aught other than a chaos of confused ruins, bewildering to those unacquainted with, or misinformed as to the conventual arrangement which anciently prevailed, with regard to the various offices or buildings which composed the monastic pile.

Monastic life is said to have originated in Egypt so early as the third century of the Christian era; and S.

Jerome, who wrote in the fourth century, informs us that there were in Egypt Cœnobites or monks, who dwelt in common, and Anchorites, who led a solitary life in the desert. In the early British Church, and up to the arrival of S. Augustine, at the close of the sixth century, there appear to have been Cœnobites, or communities of monks, who dwelt in common; but of their peculiar rule, if any, still more of their monastic arrangement, if any, we know but little.

S. Benedict, circa A.D. 516, is said to have founded, and composed the rule of that Order. It was introduced into Britain at the close of the sixth, or early in the seventh century; and the monasteries in this country of that Order seem all to have been founded between that period and the twelfth century.

The Benedictine Order was, perhaps, the most aristocratic of all the monastic Orders. The abbeys of this Order were chiefly, if not altogether, erected in or near to cities or towns. The Cistercian Order, emanating from that of the Benedictine, but of a stricter or reformed rule, founded at Citeaux in Burgundy, A.D. 1098, of which monastery S. Bernard was Abbot, was introduced into England, circa A.D. 1128, between which and A.D. 1150 most of the monasteries in England of this Order were founded. Regarded by some as farmers rather than monks, the inmates of the Cistercian Abbeys laboured in the fields, and the sites chosen for those abbeys were in secluded vales, and not in towns. Besides the great abbeys of the Benedictine and Cistercian Orders, we have in our cities and towns the remains or sites of the Friaries of the Dominican Order of the Black Friars, introduced into this country circa A.D. 1221, of which Order there are said to have been fifty-five houses; and of the Franciscan Order of the Grey Friars, introduced into this country circa A.D. 1224. Of these there are said to have been sixty-six houses. The Carmelites or White Friars, introduced here circa A.D. 1240, had about forty houses. Then we had Carthusians, a branch of the Benedictine Order, A.D. 1180, who had only nine houses, called Charter Houses; Knights Templars; Knights Hospitallers of S. John of Jerusalem; Canons of the Order of S. Augustine; White Canons of the Premonstratensian Order; of the Sempringham or Gilbertine Order, twenty-six houses; Alien Priories or Cells to foreign monasteries, one hundred and twenty in number, liable to be seized by the Crown in time of war; and Colleges of Secular Canons. Of these several Cœnobite establishments no less than six hundred were scattered over the kingdom at the time of the general suppression by King Henry VIII.

With respect to the general arrangement of monastic buildings I purpose treating only of those of the Benedictine and Cistercian Order; those of the other Orders having a general concordance, with differences according to the nature of the sites. Now there is no work I have hitherto been able to meet with which treats in a clear and distinct manner of monastic arrangement, and I have in vain searched several works on monastic discipline in a fruitless endeavour to ascertain whether any general or precise rules have been laid down with regard to such arrangement. Martene, one of the celebrated Benedictines of S. Maur, in his work, *De antiquis monachorum ritibus*, only notices, of the different conventual offices or buildings, the general relative position of one, that of the Chapter house; and even the learned Mabillon, in his voluminous work, the *Annales Benedictini*, whilst giving the measurements of a conventual church and of the monastic offices, does not treat of their relative position. Yet that there was some general and understood rule upon this subject there can be little doubt, from the general similarity of arrangement which commonly, though not invariably, prevails, for the exceptions to the general mode of arrangement are by no means unfrequent.

The sources of information we possess on this head may be collected from the ancient monastic registers or leiger books, preserved partly in our public libraries, and partly in private collections.

These frequently contain an account of each Abbot, and a notice of the buildings erected or re-constructed during his Abbacy.

In the descriptive accounts of the buildings forming some of these monastic piles, taken by the King's Commissioners on their suppression, we have, amongst others, that of the Priory of Augustine Canons at Bridlington in Yorkshire. This contains full particulars of the different monastic offices, and of their relative arrangement; though of all these, with the exception of the Gatehouse, and part of the conventual church, the eastern portion of which has been destroyed,[d] not a vestige now remains.

[d] On the suppression of the monasteries, some of the conventual

In that quaint but interesting little work, edited by J. Davies, of Kidwelly, in 1672, but drawn up soon after the suppression, entitled *The Ancient Rites and Monuments of the Monastical and Cathedral church of Durham, collected out of ancient manuscripts about the time of the Suppression*,[e] the relative position of the several offices are distinctly treated of; and plans of the monastic remains are given by Brown Willis, in his account of Durham Cathedral; in Sir H. Ellis' edition of the *Monasticon*; and by Briton, in his *Dictionary of the Architecture and Archæology of the Middle Ages*. Much information may also be gleaned from ground plans of different monastic remains, many of which have been published.

Had the two brothers, Samuel and Nathaniel Buck, who, a century and half ago, traversed the kingdom to take and engrave views of the ancient ruined abbeys, followed the plan of Kip in his representations of gentlemen's seats, and given us bird's-eye views of the different monastic remains as they then existed, their work

churches in towns were only partially destroyed, namely, the eastern portion or quire, the nave and aisles having been retained for parochial worship. Some of these had previously belonged to the parishioners, others were purchased by the parishioners from the Crown. Amongst churches still existing for parochial worship, but the eastern portions of which have been destroyed, may be enumerated the Priory church, Bridlington; S. Mary's Abbey church, Shrewsbury; Wymondham Abbey church, Norfolk; Polesworth Nunnery, Warwickshire; Waltham Abbey church, Essex; Steyning church, Sussex; Thorney Abbey church, Cambridgeshire; Pershore Abbey church, Worcestershire; besides others.

[e] In *A Description or Breife Declaration of all the Ancient Monuments, Rites and Customes belonginge or beinge within the Monastical church of Durham before the Suppression*, written in 1593, edited by the Rev. James Raine, Secretary to the Surtees' Society. Davies' work is described as being "so exceedingly rare that few people possess it."

would have proved far more valuable to us, though perhaps less artistical in plan than the kind of perspective views they have preserved. At the present day their work, as one of reference, is only of use to indicate what havoc in these crumbling piles has, since their days, taken place.

The principal monastic offices were arranged round a square or quadrangular court, sometimes called the cloister court, one side of which, the north or south, but generally the former, was bounded by the conventual church.*ƒ* In the larger monasteries there were one or two other courts, and the whole of the monastic buildings were included within a walled enclosure called the close, often containing an area of many acres, with a principal gatehouse and portals of a minor class. The walled enclosure of Boxley Abbey, Kent, with the principal gatehouse at the north-west angle, though in some parts ruinated, is continuous and unbroken.

Although there are some few churches, formerly monastic, which contain more or less portions of Anglo-Saxon architecture, as Monkswearmouth, Jarrow, Hexham, Ripon, Deerhurst, Repton, and others, I have found no such remains in the conventual buildings or offices formerly attached to such churches. The ruins of

---

*ƒ* The site of the church, north or south of the cloister court, seems in great measure to have depended on the side most convenient for the sewerage of the monastery, which was frequently carried beneath the refectory, which latter was parallel with and on the side of the court opposite to the church; but generally speaking, the site of the church, subject to the carrying out of the drainage, was on the north side. Monastic sewers were often constructed of that magnitude as to give rise to the popular but absurd opinion that they were subterraneous passages to, no one knew whither.

Innisfallen Abbey, on an island in one of the lakes of Killarney in Ireland, may, perhaps, be exceptive. These are, apparently, the earliest monastic remains I have met with, and may possibly be the original buildings of the abbey founded, according to Archdall, *Monasticon Hibernicum*, by S. Finian Lobhar, or the leper, son of Alild, King of Munster, and a disciple of S. Brendan, towards the close of the sixth century.[g] The earliest

[g] Of these primitive and remarkable ruins I took notes some six years ago whilst staying at Killarney. At p. 38 of Vol. I. of this work is an illustration shewing the rude workmanship displayed in this monastic pile, of which I may be excused for giving a somewhat detailed account. The church consists of a nave and chancel only, but the fragments of masonry which exist above ground are insufficient to shew where the one terminated, and the other commenced. The length of the church internally is 66 feet, the width 16 feet. Only one window is now apparent, this is in the south wall of the nave: it consists of a simple light covered horizontally with a lintel stone; the jambs are straight-sided and not splayed. The masonry of the church is altogether void of any mouldings, but the wall appears to have been plastered internally. On the north of the church is the court. A single wall only bounds the west side, in which side there are no appearance of buildings. The foundations of a wall, forming, as it were, a passage 5 feet 10 inches wide, appear on the south and west sides of the court.

The court measures from north to south 27 feet, and from west to east 36 feet. On the north side of the court is a plain doorway, with a horizontal lintel over, leading into one long room, apparently the refectory; the walling here is 2 feet 8 inches in thickness, and the room is about 50 feet long by 15 feet wide. At the east end is a small window, the jambs of which are partly splayed, the head of the window is horizontal. On the east side of the court are three doorways, rudely constructed. Of these the northernmost has a plain horizontal lintel; the other two exhibit semicircular-headed arches of rude construction, formed of thin laminæ of stone. Two of these open into a roofless apartment 40 feet long by 16 feet wide. The masonry is composed of irregular pieces of rock, and is very rude. Little mortar appears to have been used in the construction of the masonry, and interspersed in the walls here and there appear blocks of squared stone. At the south end of this building, high up, the only indication left of an upper story, except the pug-holes in the roof gable, is a narrow light looking into the church. This may have been the original *Camera* of the

general notice I have met with relating to monastic arrangement is that contained in the *Monasticon*, in the account of Glastonbury Abbey, where it is said, though the authority is not stated, that a new foundation was laid in the year 942, and the offices were built after a model brought from France.

Ingulphus, or the writer under that name, in giving a description of the monastery at Croyland, erected during the Abbacy of Egelricus, between the years 875 and 885, shews that they surrounded two courts, and were principally constructed of wood covered with lead.

Abbot. There are no apparent remains in these apartments of any fire-place. The walls of the court are 2 feet 6 inches in thickness. The entrance into the court is on the south-west, close to the church, where indications exist of a rude archway. Irregular blocks, larger and smaller, and small laminæ of stone form the masonry; limestone, sandstone, and puddingstone are the materials of which the walls are composed. The buildings are entirely roofless.

Separated from, and on the north-east of the main buildings, by a space of about 30 feet, is a building 33 feet in length by 15 feet in width; this appears to have been the kitchen of the abbey; and at the south-west corner are the remains of an oven.

About 42 feet west of the church are the remains of a building of block masonry, superior in construction to the original buildings, and apparently of later date. This building internally is about 36 feet long by 16 feet in width. This may have served as the Hospitium or Infirmary.

On a headland projecting into the lake and some little distance north-east of the main buildings of the abbey is a small domestic building of the twelfth century, probably the Abbot's *Camera* or lodging, and constructed some centuries after the foundation of the abbey, and in a time of comparative greater refinement. The sole apartment composing this building is internally 18 feet in length, west to east, by 11 feet 3 inches in width. On the west is a semicircular-headed doorway, enriched with the chevron and other mouldings, much weather-worn, with Norman shafts at the sides. At the east end is a small semicircular-headed window constructed of Ashlar stone; and in the north-east angle is a fire-place. The walls of this structure, which is roofless, are 3 feet 8 inches in thickness; the doorway is 2 feet 3 inches in width; and the internal splay of the window is in width 2 feet 3 inches.

He mentions amongst the buildings erected by that Abbot the infirmary of the monks, with its chapel, the guest hall *(aula hospitum)* with two great chambers, a brewhouse, a bakehouse, a granary, a stable of considerable size, with lofts over, serving as chambers for the servants of the abbey, the stable being divided into two parts, one for the horses of the Abbot, the other for those of the guests. He likewise mentions the apartments occupied by the Abbot to have consisted of a kitchen, hall, chamber, and a chapel.

He tells us also of buildings previously erected, *viz.*, the kitchen of the infirmary, the cloister, the refectory, and the dormitory; and he describes the manner in which some of the minor offices were arranged round a second court; the stable, the granary, and the bakehouse occupying the whole of the western side; the guest house, with its chambers, the south side; and the working room for the shoemaker and tailor, the hall of the novices, and the Abbot's lodgings on the east; whilst on the north was the great gate, with the eleemosynary or almonry extending from it eastward.

A valuable and well-known representation or plan, preserved in the library of Trinity College, Cambridge, of the monastery attached to the cathedral church of Canterbury, as it existed between the years 1130 and 1174, gives some idea of monastic arrangement in the twelfth century. The principal court adjoined the north side of the nave, with its aisles of the cathedral, and was surrounded by cloisters in the Norman style, with semicircular arches. I do not know of any existing example in this country of Norman cloisters, but in

## MONASTIC ARRANGEMENT. 243

Normandy, at the Abbey Blanche Mortain, is a Norman cloister supported by semicircular arches resting on cylindrical piers. This, as an existing example, may perhaps be unique. Indeed, with the exception of some gate houses and chapter houses of Norman construction, we find few remains of monastic offices in the Norman style. Exceptions indeed occur in the ambulatory or crypt-like building beneath what was the ancient dormitory adjoining the Benedictine church of S. Werburgh, now the cathedral, at Chester; at Wenlock Priory; at Castleacre Priory; and at Jarrow; but nearly all the offices of the monastic structures in this country appear to have been built or re-constructed since the twelfth century; and we find many of these subsequent erections noticed in some of the ancient monastic register books or liegers. But to return to the monastery of the cathedral at Canterbury. On the north side of the court stood the chapter house—this was the usual arrangement—but the position of the dormitory here represented on the east side, in a continuous line from the chapter house northwards, was unusual,—a departure from the general rule. The refectory occupied the proper and usual position, opposite to and in a parallel line with the church. On the west side of this cloistered court was the cellar *(cellarium)*, used as the depository for provisions. In subsequent ages we often find the cellar to be a crypt or vaulted apartment beneath the refectory. Behind the refectory, and communicating with it, was the kitchen *(coquina)*: this ranged along the eastern side of a small court, and opposite to it was the guest house *(domus hospitum)*, whilst on the south side of this court was the parlour, or monks' common room, called the *Locutorium*. On the north side was a portal, placed between the kitchen and gate house. Eastward of the dormitory was a third court, which inclosed the herbary and well of the monastery. This appears to have been surrounded by a cloistered arcade on the north, south, and east sides, and is the only instance I know of a second cloistered court; and this arcade led to the infirmary *(domus infirmorum)*, a distinct range of buildings lying eastward of this second cloistered court, and comprising a separate kitchen, chapel, and jakes. Within the close, at the north-east angle of the herbary court, was the Priors' old lodging *(camera Prioris vetus)*, whilst still further to the east stood the Prior's new lodging *(nova camera Prioris)*. The common jakes, or house of office *(necessarium)*, was a long and narrow range of building standing in the close on the north side of the second cloistered court. The bakehouse *(pistrinum)*, and the brewhouse *(bracinum)*, formed another range of buildings on this side. The granary and a bath with a chamber also formed separate buildings. The principal entrance into the close was on the west side, adjoining a structure called the new hall *(aula nova)*, probably for the accommodation of guests, and this gateway was called *porta curiæ*. Great attention appears even thus early to have been paid to the sewerage of the monastery, the lines shewing the direction of the sewers being marked on the plan.[a]

[a] A facsimile of this plan appears in the Seventh Volume of the *Archæologia Cantiana*, as one of the illustrations to the *Architectural History of the Conventual Buildings of the Monastery of Christ Church in Canterbury*, by the late Professor Willis: a paper well worthy of study.

But we must not take the account of the arrangement of the monastery of Christ Church, Canterbury, as an example of the more systematic and general arrangement which subsequently prevailed, and such as I shall have to notice in the account of that of the monastery of Durham. I may here remark that there was often a difference of arrangement in the offices of those conventual establishments situate in our cities and large towns, where the sites were not unfrequently irregular and circumscribed; and those monasteries, especially of the Cistercian Order, founded in seclusion in the country, where, the sites being ample, the buildings could be the more easily arranged on a regular and systematic plan.

The arrangement also of monastic offices which surrounded the principal court of a monastery were also subject to much change from the frequent demolition of such offices, or some of them, for the purpose of re-construction in a more costly manner, by this or that particular Abbot, in the notice of whose abbacy is often an account of what portions of the monastery he built or re-constructed.[i] Hence it is that, with the exception perhaps of the conventual church and chapter house, and in some instances of the gate house, or of some portions of the particular buildings, we rarely find monastic remains earlier than the thirteenth century.

[i] The *Register or Lieger Book of Stoneleigh Abbey, Cistercian, Warwickshire*, compiled by Abbot Thomas de Pype, elected Abbot A.D. 1352, contains up to that period the *Acta Abbatum*, or an account of each Abbot, and what was done during his abbacy. Of one of these, Peter de Wych, Abbot from A.D. 1258 to A.D. 1261, it is recorded that the only good thing he did during his abbacy was the re-building of the refectory. *Sub quo tamen constructum est novum Refectorium. Sed cum multis esset odiosus, dicebatur quod illud solum fecit bonum vid'l't quod refectorium edificavit.*

I now proceed to a summary of a general arrangement of a large monastery, including the various offices belonging to it; and the various localities of these I shall endeavour to fix, where there now exist but ruins, and those in many instances extremely slight.

The CLOSE, or walled inclosure within which the buildings of the monastery were contained, was entered through a principal GATE HOUSE, generally placed westward, or north-westward, but sometimes south-westward of the conventual church, but at some little distance from it. At S. Mary's Abbey church, York, the principal gateway was placed north-westward of the church. In the same position it occurs at Roche Abbey, Yorkshire; at Chester; at Boxley Abbey, Kent; at Malling Abbey, Kent; at Stoneleigh Abbey, Warwickshire; at Thornton Abbey, Lincolnshire; and at Lanercost Priory, Cumberland. At Bridlington, Yorkshire, the gate house lies south-westward of the church; and in the same relative position it occurs at Peterborough; at S. Albans; at the White Friars, Coventry; at Carlisle; and at Wetherall, in Cumberland.

At Furness Abbey, Lancashire, the position of the gate house was on the north side of the close, and not westward of the church: but this deviation from the general rule may be accounted for by the nature of the ground, which would not allow of its being otherwise placed.

Sometimes the gate house is the only portion of the monastery left, as at Wetherall Priory, Cumberland. Of the Priory of Kenilworth little more than the gate house remains, and this is in ruins. The same may be said with regard to the Priory of Maxstoke, Warwickshire; Roche Abbey, Yorkshire; Barking Abbey, Essex; and S. John's Abbey, Colchester.

The gate house was generally a goodly structure, and often contained chambers over the entrance gateway, being generally groined in the lower storey, having two arched entrances, one large for the entry of horsemen and carriages, and a smaller lateral entrance for those passing and re-passing on foot. At some monasteries, as at Stoneleigh, chambers and buildings were attached to the gateway, and these seem to have served as the *hospitium*, or guest house. Sometimes part of the gate house served as a prison, as that at Bridlington. This gate house is thus noticed in the Commissioners' Survey, taken at the general suppression of the monasteries:—

"The gate house: Ffurste the Priory at Bridlyngton standyth on the este parte of the towne of Brydlyngton, and at the cummyng yn of the same Priory is a gate house foure square, to toure facyon, buylded with ffree stone and well covered with leade, and one the south syde of the same gatehouse ys a Porter's lodge w$\underline{h}$ a chymney, a round stayre ledyng up to a hye chamber, wherein the three weks' courte ys always kept in w'h a chimney in the same and between the stayre foot and the same hie chamber where the courte ys kept be two proper chambers, one above the other w'h chymneys. In the north syde of the same gatehouse ys there a prison for offenders w'h in the same towre called the Kydcott. And in the same north syde ys a lyke payre of stayres ledyng up to one hye chamber in the same towre w'h a chymney. M'd that all the wyndowes of the sayd towre be clearly without glasse."

The gate house often contained a chapel, or had a chapel annexed to it, in which early service was performed for the benefit of the labourers and servants connected with the monastery. At Chertsey this chapel was described as "*Capella super portam.*" Over the gate house at Barking was a chapel of the Holy Rood. At Furness Abbey a chapel adjoined the principal gate house on the east side of it, but did not actually form part of the gate house; the dimensions of this chapel, which has been erroneously called the Abbot's chapel from the gate house being wrongly presumed to have been part of the Abbot's lodgings, is 48 feet long by 20 feet 7 inches in width. It has an entrance both from the west and south; and in the south wall, near to the site of the altar, are the remains of three sedilia. At the monastery of Pipewell was *Capella beatæ Mariæ ad portam.* At Merivale "the chappell of our Ladie near the gate of the Abby," mentioned in a deed temp. Edward III., is still existing; to this has been added, apparently since the Suppression, a nave and aisles seemingly constructed from the ruins of the old conventual church. The gate house of Malling Abbey, Kent, has a chapel attached to it on the east side; this is 23 feet long by 12 feet 6 inches in width, with a decorated window at the east end, and a two-light decorated window on the south side; the doorway is in the south wall, near the west, with a stoup on one side, and is an insertion of the fifteenth century. The roof of this chapel is high pitched, open to the rafters, and the principals have curved braces or struts, with a wind beam or collar above so as to form arches. The gate house itself is perfect, and has a greater and less entrance archway, with rooms above and on each side, and the chapel projects eastward from the north side of the gate house.

Sometimes the almery adjoined the gate house, as at Evesham Abbey; in an account of which, taken at the Suppression, is noticed a lodging or building called almery adjoining to the gate at the coming in of the said late monastery on the north.

The gate house was one of those monastic offices on the construction of which particular attention appears to have been bestowed, and it was often richly ornamented with panel-work niches and imagery on the exterior. It was also a building which suffered less from spoliation and violence than the other monastic offices, and was often left to stand entire as a habitable mansion, when the unroofing of the other principal monastic offices gradually reduced them to a state of ruin.

On entering the close through the gate house the west front of the conventual church came in view, and the monastic offices generally extending southward, and forming the west side of the principal court where there was more than one, for though in the smaller monasteries there was only one court surrounded by buildings, in the large monasteries two or even three courts were requisite for the offices and accommodation of the numerous members of the establishment, both of those immediately under the monastic rule, as also of the servants who were not, and for the guests. Entering into the principal or cloister court on the west side, through a vaulted passage under the dormitory, as at Llantony Abbey, South Wales, we find the buildings on the east side of the court extending southwards from the south transept of the church; for the larger conventual churches were generally built in the shape of a cross, with a tower between the nave and choir, and transepts issuing north and south of the tower. Another range of buildings extended east and west on the south side of the cloister court, parallel with the nave and south aisle of the church; and a returned range from south to north filled up the space to the church, along the west side of the cloister court.

Sometimes the monastic offices are found on the north side of the conventual church, but this arrangement, though not unfrequent, is an exception to the general use, and seems to have been suited to the disposition of the ground for drainage when the fall was on that side. These exceptions occur at Buildwas Abbey, Salop; at Sherbourne Abbey, Wilts; at Milton Abbey, Dorsetshire; at Tintern Abbey, Monmouthshire; at Chester Cathedral, formerly an abbey church; and elsewhere.

On the west side of the cloister court the monks' DORMITORY was usually placed, and beneath this was a passage or entrance forming a communication between the close and the court. This passage appears in general to have been under that part of the dormitory which adjoined the western portion of the church. At Llantony Abbey this entrance passage, now converted into a room, is very perfect, and is vaulted in two divisional bays with plain diagonal groins without ribs, the arch of the outer doorway being trefoil-headed, whilst the inner doorway into the court is arched segmentally.

From the only existing wall of the dormitory at Tintern Abbey we find indications of a similar arrangement; and in the remains of the conventual buildings adjoining the cathedral at Chester the same disposition is observable. This is likewise the case in the monastic buildings adjoining the cathedral at Worcester. At Buildwas Abbey the entrance passage appears to have been beneath the middle of the dormitory. In most cases, however, we find this range of building either entirely destroyed, or so altered as to give us little information.[k]

Some remains of the dormitory are, or were recently, existing near the Abbey church, Shrewsbury, but not of that portion which was immediately contiguous to the church, and which has been some time destroyed. At Fountains Abbey, a spacious groined ambulatory, divided down the centre by a row of piers, formed the substructure and now only existing remains, the dormitory over being destroyed. Remains of the walls of the dormitory likewise exist at Kirkstall Abbey; and this also had beneath it an ambulatory or sub-groining of vault work. In the remains of the White Friars monastery at Coventry, the ambulatory, with the dormitory over, still exists, though somewhat altered by partition walls. The dormitory is still, however, occupied, being divided into wards for the inmates of the workhouse, of which establishment it now forms a part, whilst the

[k] Some modern writers on conventual arrangement would place the monks' dormitory on the east side of the cloister court; this may occasionally have been the case, but I can find no reference to authority. In the monastery of Durham we have evidence of the dormitory being on the west side of the cloister court.

ambulatory beneath is used for a refectory. This, perhaps, is the most complete and interesting specimen we have left of the ancient monastic dormitory.

At the end of the dormitory adjoining the conventual church, which was generally the north end, was a flight of steps, by means of which ready access was obtained to the church, either through the cloister, or, as at Fountains Abbey, Ulverscroft Priory, Kirkstall Abbey, Worcester Cathedral, and Llantony Abbey, immediately into the church. At Wenlock Priory this access leads into a vaulted chamber within the church at the west end of the south aisle, and thence by means of a newell staircase and passage through the wall down into the church. The use of this chamber in this singular position I am unable to define, as it has no window or aperture opening into the church, but is lighted by windows opening into the cloister court. It has been suggested that this was the *Scriptorium.* Could it have been an *Anchorage?*

In general, however, the dormitory is a part of the monastery of which less remains than of many other offices. The domitory was also called the *Dorter.* Thus in the account of Bridlington Priory, where the dormitory was, contrary to the usual arrangement, on the east side of the cloister court, we find it described as follows :—" It. on the same syde of the cloyster ys the dorter goyng upp a payre of stayres of stone xx steppes highe, lying north and south, and conteynyth in length lxviij pac's and in breddyth ix pac's, also well covered with lede, and at the south end and west syde of the same dorter ys a long house of office covered with slatt."

## THE REFECTORY. 253

In the *Ancient Rites and Monuments of the Monastical Church of Durham* we find the dormitory thus described: "On the west side of the cloister there was a large house called the dorter, where the monks and novices lay, every monk having a little chamber of wainscot very close to himself, and their windows towards the cloisters, every chamber a window, by reason the partition betwixt every chamber was close wainscotted, and in every of their chambers was a desk for their books."

In the account of the goods of Pipewell Abbey, taken at the Suppression, 30th Henry VIII., we find noticed: "In the dorter the munkes selles[1] and 1 laumpe of laten."

On the south side of the cloister court, and opposite to the church, was the REFECTORY or common dining hall of the monks. Generally this extended lengthways from west to east, parallel with the church, as at Worcester; Chester; Penmon Priory, Anglesea; Shrewsbury; Lanercost Abbey, Cumberland; Finchale Priory, Durham; also at Durham; at Merivale Abbey, Warwickshire; at the ancient monastery of S. Frideswyde, Oxford; and at S. Mary's Abbey, York. The dais or upper part of the hall was at the east end. At Fountains Abbey, however, and at Tintern, and at other Cistercian abbeys, the arrangement of the refectory seems to have been somewhat

---

[1] Anciently there were no divisions in the dormitory, but the beds of the monks were exposed. The Synod of Exeter, held A.D. 1287, treats of this practice, cap. xi. De dormitorio. *Cellas in dormitorio esse, Benedictus papa duodecimus prohibuit; nosque hoc adjiciendum decernimus, quod ad tollatur omnis mali suspicio, lecti monachorum velaminibus, si quæ fuerint, amotis, et perticis ita sint ordinati, ut in ipsis lectis existentes vel juxta, sine obstaculo quocunque die noctuque continue valeant a custodibus ordinis, et aliis transeuntibus intueri.—Wilkins' Concilia.*

different, the refectory extending lengthways from the court north to south. Whether this arrangement was original, or whether it followed on a re-construction, is a difficult matter to determine, as we have not remains sufficient to enable us to judge.

The refectory contained a pulpit, in which certain portions of Scripture were read during the repast. In the *Rites of Durham* the mode of reading is thus noticed: "One of the novices at the election and appointment of the master, did read some part of the Old and New Testament, in Latine, at dinner time, having a convenient place at the south end of the high table within a fair glass window compassed with iron and certain steps of stone, with iron rails on the one side to go up to it, to an iron desk there on which lay the Holy Bible."

These pulpits were frequently of stone. An early and well-known specimen remains at Beaulieu in Hampshire, the refectory of the abbey of which now serves as the parish church. This is of the thirteenth century. Another fine specimen of the same period is well preserved in the refectory, Chester Cathedral, formerly an abbey church. A fine specimen of the Decorated style exists in the pulpit, now almost the only remains of the refectory, S. Mary's Abbey, Shrewsbury. At Tapholme, in Lincolnshire, the pulpit of the refectory is also remaining. The passage through the walls and steps up to the recess projecting outwards for the pulpit is existing in the south wall of the refectory, now in ruins, of Merivale Abbey, Warwickshire; and the recess, with a window at the back, for the pulpit in the remains of the refectory of Ulverscroft Priory, Leicestershire, is still visibly apparent, though in a very ruinous condition. The general position of the pulpit was in or projecting from the south wall; but where, as at Tintern, the refectory ran lengthways from north to south, we also find remains of the passage leading to it in the west wall; and we also find it in the same position at Kirkstall.

At Pipewell Abbey the goods of the refectory are thus noticed in the *Survey* taken at the Suppression: "The Frater.—It ther 3 bordes 1 pulpytt 11 tables 2 payr of truseulles 1 forme, sould 0 „ 2 „ 0."

Under the refectory we sometimes find a vaulted apartment or crypt, which served as the cellar or depository for provisions. The groined roof of this apartment was occasionally supported by a row of piers, as at Lanercost Abbey, Cumberland, and Finchale Priory, Durham. The refectory at Penmon Priory, Anglesea, appears to have had a cellar beneath it, but there are no indications of the latter having been vaulted or groined; this cellar is lighted on the south side by oblong apertures. Of the few existing remains of the monastery attached to the Cathedral church, Carlisle, is the ancient refectory, now converted into a dwelling-house. The ancient refectory of the monastery of S. Frideswyde, Oxford, is still existing, now I believe converted into rooms; but on the south side of the exterior may be seen the projection within which the pulpit anciently was placed. In the monastic buildings adjoining the cathedrals of Chester and of Worcester the ancient refectories are used as school-rooms. The refectory was ofttimes called the *Fratry*, as in the notice of the Priory of Bridlington, " It' on the south syde of the same cloyster ys the Fratre, which conteynyth in length xxiij pac's and in breddyth x pac's, buylded w'h free stone, and well covered with lede." In the *Rites of Durham*, this building, of which there are still considerable remains, though much altered, is particularised at length: " In the south alley of the cloisters is a fair large hall, called the *Frater House*, finely wainscotted on the north and south sides, as also on the west. And on either part of the Frater house there is a fair long bench of stone mason-work, from the cellar door to the pantry or cover door. Above the bench is wainscot-work two yards and a half in height, finely carved and set with embroidered-work; and above the wainscot there was a fair large picture of our Saviour Christ, the blessed Virgin Mary, and Saint John, in fine gilt-work and excellent colours; which pictures, though washed over with lime, yet do appear through it. This wainscot-work hath engraven on the top of it, *Thomas Castell, Prior, Anno Domini* 1518, *mensis Julii*. So that Prior Castell wainscotted the Frater house round about."

Near the entrance from the cloister court into the refectory was placed the lavatory, an oblong trough of stone cased with lead, at which the monks were accustomed to wash before dinner. In the compotus for Finchale Priory for 1367-8 is an entry, "*pro lotoriis factis in claustro xxxvii<sup>s</sup>*." In an inventory of the treasure of the church of Peterborough, taken A.D. 1539, is the following:—" In the cloyster, Item one conduit or lavatory of tynne with divers coffers and seats there."

The lavatory is thus mentioned in the *Rites of Durham*: "Within the Cloyster Garth, over against the Frater House dour, was a fair Laver or Connditt, for the Monncks to washe their hands and faces at, being made in forme round, covered with lead, and all of marble, saving the verie uttermost walls. Within the which walls you may walke round about the laver of marble, having many little cunditts or spouts of brasse, with xxiiij cockes of brasse rownd about yt, having in yt vij faire wyndowes of stone woorke, and in the top of it a faire Dove-Cotte, covered fynly over above with lead, the workmanship both fine and costly, as is apparent till this daie. And, adjoyninge to the est syde of the Counditt dour, ther did hing a bell to geve warning, at a leaven of the clock for the Monncks to cumme wash and dyne, having ther closetts or almeries on either syde of the Frater House dour keapt alwaies with swete and clene towels, as is aforesaid, to drie ther hands." In the ancient chorister's vestry in the eastern south transept, Lincoln Cathedral, is a stone lavatory panelled in form of the Decorated period. In the cloister adjoining Norwich Cathedral are lavatories of the fifteenth century. Remains of the lavatory in the centre of the cloister garth, Durham Cathedral, are still existing. At Salisbury Cathedral the ancient stone lavatory of early fifteenth century work, though still existing does not occupy its original position. The stone lavatories in the north cloister of Gloucester Cathedral—the monastic offices being northward of the conventual church—are perfect. Other instances might be adduced.

Near to or adjoining the refectory, either on the east or west side, or behind it, was the great kitchen, *coquina*, of the monastery. That at Fountains Abbey is, perhaps, the most perfect specimen we have: it is divided into two compartments, with two immense fire-places, the arches of which are constructed with the somewhat rare description of masonry termed "juggled." The kitchen of the monastery annexed to Chester Cathedral, the monastic offices of which were on the north side of the conventual church, was behind the refectory, and adjoined it on the north-west angle. In the remains of Tintern Abbey a room adjoins the refectory on the west side; and in the wall which divides it from the refectory is an opening like a Buttery hatch, which it probably was. The Abbot's kitchen at Glastonbury appears to be the only portion of the monastic offices now remaining; it is exceedingly curious, being octagonal in shape, with a roof like that of an octagonal-shaped chapter house, with a lantern in the centre; but it is now difficult to ascertain the precise position this had with regard to the other monastic offices, as it lies considerably to the south-west of the site of the cloister court. It may have been merely the kitchen attached or belonging to the Abbot's lodgings.

At the south-east corner of the cloister court, or at the corresponding or north-east corner, where the monastic offices were on the north side of the conventual church, we ofttimes find the vaulted substructure or remains of a range of buildings extending lengthways from north to south, and forming the Abbot's or Prior's lodgings, *Camera Prioris*. In this position are the remains or vaulted cellars of the Abbot's lodgings at Kirkstall and Stoneleigh, where the arrangement is very similar. Sometimes instead of a single range we find the remains of an irregular group of buildings at the south-east angle of the cloister court, and adjoining it. There were, however, exceptions to this general plan of position, as at Bridlington and Castle Acre Priory, where the Prior's lodgings appear to have occupied the north-*west* angle of the cloister court. In large monasteries the Abbot's lodgings contained a reception room or hall, chambers, kitchen, cellars and other offices, including a private chapel or oratory. The Prior's lodgings at Wenlock Priory are more perfect than we usually find them ; they form a range of building constituting the eastern side of a second court, at the south-east angle of the cloister or principal court. On the ground floor at the north end is a small chapel or oratory,[m] still containing within a projecting window recess the ancient stone altar, which is not plain but panelled in front; and it is the only instance of an ancient panelled altar I have met with[n]; and in the south wall near the altar is a piscina. The kitchen and other domestic offices are on the ground floor; and along the western side is a curious corridor or passage, forming, as it were, a kind of cloister, with a series of panel-like windows formed of stone-work, but unglazed, running along the whole range. The floor above is ap-

[m] In a Charter of Finchale Priory, dated 1474, the private chapel of the Prior is thus noticed: "Capella Sancti Nicholai juxta cameram domini Prioris situata."

[n] I have an impression of a monastic seal on which is represented an altar with a panelled front.

proached by a flight of steps, which leads into a narrow corridor or passage similar to that beneath, the windows or openings along the front being formed of open and closed panel-work divisions. From this upper corridor is entered what appears to have been the Prior's hall or reception room, having a high-pitched roof with curved principals open to the rafters.

This room is lighted on the east side with four windows of two lights each recessed in the wall. In the recesses, which are panelled, are stone shelves. The south wall of this room exhibits traces of mural paintings. South of this hall is a chamber, entered from the corridor, which contains a fire-place in the south wall, and is lighted on each side of the fire-place by a window, and by two other windows. At the north-east corner of this chamber is a perforated stone basin, which appears to have been used as a sink, there being a projecting gargoyle on the outside. In the east wall there is a door, with a flight of steps leading downwards to a closet, apparently a jakes. In the north wall at the south-west corner is another staircase, which appears to have constituted a direct communication with the kitchen and offices below. This range of building is deserving of a careful plan and more detailed description than that I have now attempted. At Bridlington the Prior's lodgings occupied the position generally assigned to the dormitory, along the west side of the cloister court, and adjoining the church, and are thus described: "There standith on the south syde of the said church the Prior's lodgyng, wherein ys a hawle to the whiche hall ledyth a stayre of iiij foote brode and of xx steppys highe, which stayres be on the south syde of the same hall; the seyd hall conteyneth in length from the Skyven to the high desk xviij pac's, and in breddyth x pac's and well covered with lede. It' on the north syde of the same hall ys there a great chamber where the Priour alwayes dyned, conteyning in length xx pac's, and in breddyth ix pac's well covered with lede. It at the west ende of the same great chamber ys there a proper ytle chamber, whiche was the Prior's slepyng chamber, covered with lede: and ov' the same chamber ys a garratt. It. at the est syde of the same great chamber ys a lytle chappell with a closett adjoynyng to the same. It. at the south ende of the hawle is the buttrie and pantrie under one office, and one the same ende a chamber called the audytor's chamber. It. at the same ende of the hawle and on the west syde ys a fayre plo'r or a chamber called the lowe som' parlo'r ov' the which som' parlor or chamber ys another fayre chamber covered w'h lede, and adjoynyng to the same highe chamber on the est side be three lytle chambers for servaunts. It. at the south ende of the same hawle ys the Pryor's kechyn, whiche ys an olde kechyn, with three lovers covered with lede, and adjoynyng to the same kechyn ys there a chamber called the south sellerer's chamber."

Although not a trace now remains of the buildings thus described, the full notice of them in detail is useful as tending to shew what offices constituted the Prior's Lodge. The apparent remains of the Abbot's lodgings at Tintern extend from the north-east corner of the court northwards. They comprise a kind of ante room and a range of building 78 feet long by 25 feet wide, with a row of piers, five in number, down the centre; the substructure therefore appears to have been vaulted as elsewhere. The remains of the Prior's lodgings at Ulverscroft Priory, Leicestershire, extended eastward in a line with the refectory, at the south-east angle of the court. These are now converted into a farmhouse. Sometimes the Abbot's or Prior's lodgings constituted a distinct building or dwelling, and formed part of a second court, as at Finchale Priory, Durham, where a long range of building with vaulted cellars constitute the Prior's lodgings, and are parallel with the choir of the church, forming the south side of a second court.

CHAPTER HOUSE. On the east side of the cloister court, and generally separated from the north or south transept of the conventual church by a narrow oblong apartment or passage, we almost invariably find the chapter house, where the monks, with the Abbot or Prior at their head, assembled in conclave. This particular position is assigned to the chapter house by the learned Martene, in his work, *De Antiquis Monachorum Ritibus*: "*Locus capituli erat claustro contiguus ad orientem.*" Up to the close of the twelfth century, and during the time those styles of architecture were prevalent which we call Norman and semi-Norman, the plan of the chapter house was that of a square or parallelogram, as at S. Mary's Abbey, York; Buildwas Abbey, Salop; Wenlock Priory, Salop; and Stoneleigh Abbey, Warwickshire. In some of the small conventual houses, as at Wroxhall Priory, Warwickshire, this plan was continued to a much later period. The rectangular-shaped chapter house also appears at Llantony and Tintern. The entrance to a Norman chapter house was through a large semicircular-headed doorway, often deeply recessed, and much enriched with ornamental mouldings; and this doorway was flanked on each side by a semicircular-headed window, divided by a shaft into two semicircular-headed lights. This arrangement in front of a Norman chapter house we find to exist at Bristol Cathedral; at Combe Abbey, Warwickshire; and at Haughmond Abbey, Salop, a late specimen in the semi-Norman style. In small monasteries the front of the chapter house, if Norman, was simply broken by a semicircular-headed doorway, without the windows I have described as flanking it on each side. Sometimes there was an apartment or ante-room between the entrance to the chapter house and the chapter house, as at Bristol and Chester.

From the commencement of the thirteenth century the chapter house of a large conventual establishment was often of an octagonal or polygonal shape, supported in the centre by a pier, with a stone vaulted roof.....Round the sides of the chapter house was a stone bench, on which the monks sat whilst in chapter. At the external angles of these polygonal chapter houses flying buttresses were often disposed, so as to counteract the pressure or thrust on the walls of the stone vaulted roof, whilst each side of the wall was pierced by a window of large and elegant dimensions, filled with geometrical or flowing tracery. Such chapter houses in a perfect state are annexed to many of our cathedrals, as those of Salisbury and Lincoln, and of the Abbey church, Westminster, which are of the thirteenth century, and those at York and Wells, which are of the fourteenth century.

We sometimes find amongst the ruins of conventual buildings, not cathedrals, traces of the polygonal-shaped chapter house. Considerable remains exist of that at Howden in Yorkshire; this is a structure of the fifteenth century. The foundations, or rather basement walls, of the polygonal chapter house of the Priory at Kenilworth were discovered a few years ago, having been concealed probably soon after its demolition, on the Suppression, by heaps of rubbish. The chapter house at Bridlington Priory was of a polygonal shape, as appears from the account taken at the Suppression: "It. on the est syde of the same cloyster ys a very fayre Chapter House wh ix fayre lights about the same wh whyte glass and sume imagerie, covered wh lede spere facyon." The entrance to the chapter house from the cloister court was near a doorway in the wall of that aisle of the church contiguous to the cloister and near the east end, so that the monks could easily pass in procession from the chapter house into the church.

MORTUARY CHAMBER. Between the chapter house and the north or south transept, as it might be of the conventual church, there was generally a passage or narrow apartment, the exact name of which I have been unable to ascertain, though in the *Rites of Durham*, at which monastery it was used as a passage from the cloister to the cemetery garth, it is called *a parlour*. "And after there devocion the dead corpes was caryed by the Monnckes from the Chapter house thorough the Parler, a place for Marchannts to utter ther waires, standing betwixt the Chapter house and the church dour, and so throwghe the said parler into the Sentuarie Garth, where he was buryed." At S. Alban's Abbey this apartment is a kind of closed cloister or passage of Norman construction, very curious, and it is entered from a doorway through the south wall of the south transept of the church. With the exception of the abbey gate house this passage appears to have been the only remains of the conventual buildings attached to or adjoining the abbey church of this once famous monastery. At Combe Abbey the doorway to this passage or room appears on the east side of the cloister court. We find the remains of this narrow apartment at Tintern Abbey and at Llantony. At Tintern the space between the north transept and chapter house is ten feet wide; and this is divided by a cross wall. The eastern portion of this space appears to have been vaulted and used as a revestry, there being a doorway into it from the transept. The western portion formed an apartment ten feet wide and about twenty-five feet long, entered from the cloister through a double doorway. At Llantony Abbey the space between the south transept and chapter house is about ten feet wide, exclusive of walls, and is vaulted in two bays, forming an apartment twenty feet long from west to east, the entrance being through a pointed arched doorway in the west wall. This apartment, *slypp*, or passage, is so general a feature that the instances where it does not occur are rare. Exceptions, as at Furness Abbey, where the chapter house immediately adjoins the south transept of the church. At Wenlock Priory also the chapter house adjoins the south transept of the church, there being no intervening passage or apartment between them; but we have here a singular arrangement, for along the west wall of the south transept is a vaulted slypp or passage, nine feet wide, and groined lengthwise in three bays or divisions, a difference of arrangement I have not met with elsewhere. Sometimes this narrow apartment is entered immediately from the church, as at Fountains Abbey and at Kirkstall, and may have served as a revestry. By some, and not without reason, this slypp or passage is considered to have been a mortuary chamber or receptacle for the bodies of dead monks before burial.

Of the conventual church, the south or north aisle of the nave of which generally formed the boundary, north or south, of the cloister court, I purpose presently to treat.

Along each side of the principal court was perhaps generally, but not always, a covered ambulatory or cloister, the roof of which was ofttimes groined in stone with numerous vaulting ribs, according to the particular architectural style in which it was constructed, with large windows in the walls towards the court filled with tracery, geometrical, flowing or panelwork. We find the cloister as an appendage to a monastery early mentioned in the eleventh and twelfth centuries, as in the year 1173, in the chronicle of Jocelin of Brakelond, though I know of no specimen of so early a period now existing in this country. Cloisters of the thirteenth century exist at Salisbury; and those at Peterborough were of the same date. Remains of cloisters in the Early English style also exist at Malling Abbey, Kent. At Norwich Cathedral they are of the fourteenth century; and as these were built progressively, through nearly the whole of that century, they exhibit transitional features, though as to style they are not of a transitional character, being of that usually known as the "Decorated," but they are very interesting. We do not, however, often meet with cloisters in this country of an earlier date than the fifteenth century, examples of which period are numerous.

In some of the smaller conventual houses we find no indication of cloisters, as at Wroxhall and Stoneleigh in Warwickshire; and in many instances we have reason to infer that the cloister was an appendage or adjunct, perhaps originally intended, but constructed long subsequent to the original building, as that at Combe Abbey, Warwickshire.

Such then was the general arrangement of the principal offices of a monastery situate with regard to the position of the church. The chapter house on the east side of the cloister court, forming a continuation of the buildings on a line with the transept, north or south, of the church, the slypp or passage or narrow apartment I have noticed only intervening between them. On a line with the chapter house, north or south, were the Abbot's or Prior's lodgings extending beyond the cloister court. The refectory, and in many cases vaulted cellar beneath, with the kitchen near adjoining, occupying the side of the cloister opposite to and parallel with the church, and the dormitory with, occasionally, a vaulted ambulatory beneath, and passage near the church from the close into the cloister court occupying the west side of the court, the cloister being continued all round. The court itself was used as a cemetery garth, or burial place for the monks, whilst in the chapter house were inhumed the remains of the Abbots or Priors, though these were also not unfrequently buried in the choir of the monastic church.

But there were also minor offices and rooms sometimes forming part of the buildings round the cloister court, and sometimes disposed elsewhere. Of these was the LOCUTORIUM, parlour or common room of the monks.

This at S. Mary's Abbey, York, was adjoining the refectory, being a continuation of the buildings on the south side of the court, and eastward of the refectory. At Fountains Abbey it was also on the south side of the court, but adjoined the refectory on the west; and at Durham it was going out on the west side of the cloister.

The position of the SCRIPTORIUM, where manuscripts were written, copied, and illuminated, I have no means of defining.

At some of the monasteries, as at Durham, Gloucester, and Chester, the monks were accustomed to study or read in carrels, or small inclosed closets of wainscot in the cloisters, a portion of which was so designed and contrived purposely to admit of these. Near to the carrels in the cloisters at Durham were certain ambries or closets, which contained the monastic library, for there does not appear to have been any room in the monastic offices specially applied to that purpose; and in the will of William Place, Master of S. John's Hospital, Bury St. Edmunds, made in 1504, he bequeaths "to the Monastery of Seynt Edmund forseid my book of the dowts of Holy Scryptur, to ly and remain in the cloyster."

Sometimes, however, we find special buildings appended to monasteries as libraries. The library, with other parts of the Priory of White Friars, London, were granted 32 Henry VIII., A.D. 1541, to Richard Moresyne. Archdale informs us that the library of the Cistercian Abbey of Newry, County of Down, Ireland, was, with all its effects, and a yew tree belonging to it, planted by the hands of S. Patrick, destroyed by fire A.D. 1162. Richard Whittington, A.D. 1429, founded the library of the Grey Friars of London; it was 129 feet in length, and 31 in breadth, all wainscotted about, having twenty-eight desks, and eight double settles of wainscot; this was finished in the following year, and within three years after furnished with books at the expense of £556. 10s. 0d., whereof Richard Whittington gave £400., and Dr. Thomas Winchelsey, a friar there, gave the other £156. 10s. 0d.; and for the transcribing of the works of Nicholas de Lyra in two volumes, to be chained there, 100 marks. The library of the Priory of Norwich, "Bibliotheca Christicolarum Nordovice," is mentioned by Leland.

In the Priory of Depyng in Lincolnshire, a cell to the Abbey of Thorney, was a library, the catalogue of which has been preserved, thus headed, "Isti sunt Libri de Armariolo Monochorum de Est depyng."

In the Bodleian Library is a MS. containing a catalogue of the library belonging to the Nunnery of S.

Martin, Dover, thus styled: "Præsens hæc matricula bibliothecæ prioratus Dovorræ anno incarnationis Dominicæ 1389." The library of the Abbey of S. John, Colchester, is mentioned by Leland.

At Gloucester Cathedral, that which was the abbey library was on the foundation of the chapter, converted into the college school. In the *Rites of the Monastical Church of Durham*, we are informed that "there was a Lybrarie in the south angle of the Lantren, which is now above the clock, standing betwixt the Chapter House and the *Te Deum* Wyndowe being well replenished with ould written Docters and other histories and ecclesiastical writers."

TREASURY HOUSE. The treasury house was at Durham on the west side of the cloisters, close to the door leading up to the dormitory. At Bridlington the treasury house was also near the dormitory.

INFIRMARY. The infirmary was either a separate building, or formed part of a range of buildings round a second court, and it often contained a separate refectory and a chapel. No precise general local position can be assigned to it, but its place seems to have been dictated by mere convenience. At Worcester the infirmary stood westward of the dormitory; and the same was the position of the infirmary at Durham.

HOSPITIUM. The hospitium, domus hospitum, or guest house, for the reception of strangers, also appears to have had no general local position, but sometimes formed a separate building, and sometimes part of a range, and in the large monasteries comprised the guesten or guest hall, which served as a refectory, and several chambers for the accommodation of strangers, and sometimes a chapel. The remains of certain detached buildings at Furness Abbey, supposed to be those of the hospitium, (though this is a conjectural opinion), consist of a chapel with a chamber over it; the foundations of a considerable range of buildings westward to this have been traced. These appear to have abutted on the southern boundary of the close, but they may have been remains of the infirmary.

At Worcester the guesten hall was remaining in a comparatively perfect state until within a few years, when it was destroyed, for what reason I know not. The hospitium was sometimes in the group of buildings in which was the gate house, as at Malling Abbey, Kent, and Stoneleigh Abbey, Warwickshire; but these instances are conjectural only. The guest house at Durham is thus noticed in *The Rites*. "There was a famouse house of hospitallitie, called the Geste Halle, within the Abbey garth of Durham, on the weste syde, towards the water, the Terrer of the house being master thereof as one appoynted to geve intertaynment to all staits, both noble, gentle, and what degree soever that came thither as strangers, ther intertaynment not being inferior to any place in Ingland, both for the goodness of ther diett, the sweete and daintie furneture of there lodgings, and generally all things necessarie for travellers. This haule is a goodly brave place, much like unto the body of a Church, with verey fair pillers supporting yt on ether syde, and in the mydest of the haule a most large rannge for the fyer. The chambers and lodginges belonging to yt weare swetly keept, and so richly furnyshed that they weare not unpleasant to ly in, especially one chamber called the Kyngs Chamber, deservinge that name, in that the King himselfe myght verie well have lyne in yt for the princelynes thereof."

BAKEHOUSE AND BREWHOUSE. *Pistrinum et Bracinum* generally adjoined each other; sometimes they formed a separate building, but sometimes, as at Chester, they formed part of a range round a second court.

THE ABBEY MILL. This, an important adjunct, stood within a short distance of the monastery, on the stream near to which the monastery was generally placed. Sometimes, in particular situations where there was no stream, a horse mill was used in its stead, as at Bridlington. "It. on the north syde of the same Bakehouse and Brewhouse standyth a ffayre Horse Mylne newly bylded and covered w'h slatt."

There were other buildings, especially at the large monasteries, as stables, granaries, barns, and other farm offices, but arranged in no definite position with regard to the monastic structure. A little to the south-west of the site of Boxley Abbey, Kent, is an immense grange or range of building, 150 feet long, which appears to have served for many, if not all, of the above offices. A small grange or barn is likewise standing to the south-west of the site of the Priory at Kenilworth; but in the smaller monasteries there appears little doubt but that many of the minor offices were dispensed with.

The SEWERS of the monasteries were good, and in sanatory matters these structures were far in advance of the ages in which they were constructed. The position of a monastery was in general fixed close to a river or flowing stream, into which the sewers were carried, ofttimes under the refectory. At Tintern a sewer is carried under the remains of the Abbot's lodgings. At Malling, Kent, eastward of the site of the church is a very deep and large sewer. The size, indeed, of some of these sewers has evidently given rise to those stories often prevalent of subterraneous passage of unknown but immense length, leading from this or that particular monastery to indeed we know not where.

THE CONVENTUAL CHURCH. Having thus far attempted to give a summary of the various monastic offices and their relative arrangement, I shall now endeavour to give a description of the conventual church, and shew in what respects it differed generally from a mere parochial church. Thus we find the west front to have contained the principal entrance; and this appears to have been accessible to all who entered the close or inclosed area within which the monastic buildings stood. From the cloister court into the south aisle, as it might be, of the nave of the conventual church, were two entrances through the south wall, one near the west, the other near the east end, both opening into that side of the cloister abutting against the church. The westernmost of these entrances appears to have served for the ingress of the monks from the dormitory through the cloister into the church at the nocturnal offices. Sometimes access was immediate from the dormitory into the church, without descending into the cloister; and we find steps leading down into it direct from the dormitory. Arrangements of this kind appear at Llantony Abbey; at Wenlock Priory; at Ulverscroft Priory;

at Fountains Abbey; and at Worcester Cathedral. The easternmost entrance from the cloister was either one of egress from the church in procession to the chapter house, or more particularly for the use of the Abbot or Prior; but authority as to the object of this particular entrance, which is only, I believe, to be found in a conventual church, has still to be sought for.

The conventual church was generally built in the form of a cross, having a nave with aisles, a central tower between the nave and choir, and transepts north and south of the tower, whilst eastward of the tower was the choir. Beyond the choir we sometimes find the presbytery with aisles, often an after annexation; and in some conventual churches eastward of the presbytery was the Lady chapel, or chapel dedicated specially in honour of the Blessed Virgin, as at St. Albans. Sometimes, however, the lady chapel was erected by the side of the choir, and not eastward of it, as at Oxford and Ely. In general design the conventual churches of the Benedictine and Cistercian Orders materially differed. Those of the latter Order were extremely plain, with a shallow choir, the east wall of each transept being divided into two or three compartments, in each of which was an altar, but no separate chapels; whilst the conventual churches of the Benedictine Order, built with regard to greater magnificence, had an elongated choir, the eastern termination being often apsidal or polygonal, with apsidal or polygonal chapels surrounding the choir, and placed also eastward of the transepts. The conventual churches of Canterbury, Peterborough, Norwich, Gloucester, Tewkesbury, and Romsey, were originally thus disposed. In the conventual churches, however, re-built or added to in the thirteenth, fourteenth, and fifteenth centuries, we often find the transepts with one aisle each ranging along the east sides. This arrangement is very clear at Kirkstall Abbey church. Like indications are also to be met with in the ruins of Roche Abbey church; these aisles, and a small portion of the choir, being all that remains of that church. At Furness Abbey church, in the north transept, is a similar arrangement; and the same arrangement likewise occurs at Whitby Abbey church.

The choir was entered from beneath the tower, under the rood-loft gallery, the screen of which in cathedral and some other conventual churches was sometimes of stone, of rich architectural design, with niches containing sculptured imagery. On entering the choir we find on each side a range of stalls on the west, returning north and south. Of these, those in our cathedral and in some of our parish churches formerly collegiate or belonging to some monastery, as at Higham Ferrers, Northamptonshire, and Astley and Knowle, Warwickshire, still remain. These stalls had a continuous desk before them, and were disposed on each side of the entrance facing the east, and thence returning at right angles, continued eastward to a certain distance down the western portion of the choir. The seats of these stalls were moveable, that is to say, they were so disposed that at certain periods of the services they were turned up, and a small ledge presented itself, on which the monk who occupied the stall reclined rather than sat.

These moveable seats—I have before treated of them

—were called MISERERES or misericords; and beneath them, on the SUBSELLIA, we find those quaint and fantastic carvings which have excited much attention: for many are of a satirical cast, aimed against the Friars preachers; some are designed to teach us not to slight our mortality, as the Dance of Death carved on the subsellia of some stalls in S. Michael's church, Coventry; some illustrative of medieval history and romance; some even of pagan mythology. A long and discursive range of subjects might on inquiry be found, to which these carvings are allusive. These stalls were in cathedral and large conventual churches richly canopied. Perhaps the most beautiful and chaste, though not the most richly designed canopies, are those over the stalls at Winchester Cathedral; these are of the fourteenth century. In small conventual churches we not unfrequently find the stalls without canopies.

THE HIGH ALTAR. At the east end of the choir, and sometimes, though often not, of the conventual church, was placed the high altar, with, on the south side, the canopied sedilia for the priest, deacon, and subdeacon; and eastward to these, and within a richly canopied fenestella or niche was the piscina, sometimes with two perforated basins, the uses of which I have alluded to. At the back of the high altar was often, especially in the fifteenth century, erected a gorgeous screen or reredos of tabernacle work, of rich architectural design, with imagery in canopied niches, and sometimes sculptured tables in bas-relief.[*] The altar

[*] In the *Compotus* for the Priory of Finchale for the year 1376-7 is an entry, "*In solucione facta ad le reryrerdos magni altaris L*.*"

screen or reredos, St. Albans Cathedral, completed by Abbot Wallingford, A.D. 1480, exhibits a mass of detail in canopied niches and tabernacle work, from which the imagery has been removed. In the Museum at Taunton are the remains, skilfully put together, of a sculptured reredos dug up in the chancel of S. John's church, Wellington, Somerset. This is a rich and most interesting specimen of the fourteenth century, and consists of three principal compartments with imagery, as the Crucifixion, with S. Mary and S. John; S. Michael weighing souls; S. Christopher; and others.

The choir, like the nave, often had aisles, shut out by the screenwork at the back of the stalls between the arches, and by a parclose beyond, whilst the high altar was often set against an insulated wall, behind which was sometimes a revestry. Beyond the choir was the presbytery or retro-choir, comprising the space between the high altar and the east end of the church, or the high altar and the lady chapel. When, as was often the case, the conventual church contained the enshrined relics of some canonized or reputed saint, held in more than mere local estimation, to whose shrine pilgrimages were made, this part of the conventual church, that is, the presbytery, usually contained the stone shrine of rich tabernacle work which inclosed the *feretrum* or wooden coffer, in which the relics were more immediately deposited.

This was the position of the shrine of S. Edmund at Bury; of that of S. Cuthbert at Durham; at St. Albans, with the shrines of S. Alban and S. Amphibalus; at Westminster, with that of King Edward the Confessor.

The position of some other shrines there may be a difficulty in tracing, as in the shrines of S. John of Beverly at Beverly; of S. Wendreda at Ely; of S. Rumon at Tavistock; of S. Chad at Lichfield; of S. Frideswyde at Oxford; of S. Werburgh at Chester; of S. Wenefred at Shrewsbury. The remains of the stonework of the shrine of S. Amphibalus has within the last few years been discovered, and the numerous fragments joined together, the whole, though imperfect, exhibiting one of the most beautiful and costly monuments we possess of the fourteenth century. The shrine of S. Edward the Confessor is still existing at Westminster; and that of S. Ethelbald in Hereford Cathedral. Fragments of the shrine of S. Werburgh at Chester have been worked up; and a portion of the shrine of S. Wenefred is still remaining in the Abbey church, Shrewsbury. What I conceive to be the shrine of S. Dunstan at Canterbury Cathedral, and in which his remains may still be inclosed, is the stone structure which has long been treated as the monument of Archbishop Theobald.*

<sub>*</sub> S. Dunstan died A.D. 988. Early in the sixteenth century, in consequence of the monks of Glastonbury having laid claim to the custody of his bodily remains, which they asserted were in their possession, a search took place in the reputed shrine of that saint in Canterbury Cathedral, on the 22nd of April, A.D. 1508, Warham being then Archbishop, and Golston Prior. The result of this investigation was recorded in a document drawn up at the time, and published by Henry Wharton in his *Anglia Sacra*, A.D. 1691. This document is entitled *Scrutineum factum circa feretrum beatissimi Patris Dunstani Archiepisc*, &c. The monks employed in the search at dead of night are reported to have found a certain leaden cist or coffin in which the relics of the saint were inclosed, which cist was deposited in a stone shrine. *Arcam quandam plumbeam, ubi sacræ reliquiæ recondebantur Quæ quidem arca deposita fuit et immersa in opere lapideo feretri.* Within this was a wooden cist, covered within and without with lead, and nailed in every part, so that between nail and nail there was no space left of the

As these shrines were enriched by the votaries who resorted to them with many valuable gifts and offerings, monks were appointed whose special office it was to

width of the palm of one's hand. *Ea siquidem arca intus erat lignea, exterius interiusque plumbo undique cooperta, et clavis omni in loco affixa; adeo ut inter clavum et clavum non erat spacium relictum latitudinis humanæ palmæ.* This cist was also nigh the length of the stonework of the same shrine, viz., vii feet. The width of the cist was a foot-and-half, and it was in every part bound round with iron bands so securely that there appeared hardly any possible way of opening it. *Erat quoque hæc arca longitudinis juxta longitudinem ipsius feretri, videlicet vii pedum. Latitudinis circiter pedis cum dimidio. Eratque in omni suâ parte ferreis ligamentis circumducta tutissimè, adeo ut vix possit discerni via possibilis illam aperiendi.* Six of the monks deputed for this purpose by the Prior, with the help of others, with vast labour caused this cist of great weight to be lifted above the stonework of the shrine. *Sex enim de confratribus per Priorem ad hoc deputati una cum ope aliorum (quos convocarunt) ingenti sudore hanc arcam, quæ est magni ponderis, fecerunt supra opus lapideum sublevari.* When that was done, they at length, with great difficulty, busily employing themselves, opened the front of the cist. *Id cum fecissent, tandem cum magna difficultate satagentes anteriorem partem arcæ aperiunt.* Which they were unable to do effectually, lest they should break part of the wood with which the cist was shut in on the upper part of the cist. *Quod profecto facere nequissent, nisi partem asseris, quo in superiore parte arca claudebatur, effringerent.* That being skilfully broken, it was permitted to see within from one end of the cist to the other end. *Eo sane confracto, licuit videre interius est ab uno fine arcæ usque ad alterum finem.* There indeed lay open to the sight a certain cist of lead, which cist was constructed not of plain lead, but skilfully folded together. *Ibi vero patebat aspectui cista quædam plumbea: Quæ quidem cista facta est non ex plano plumbo sed arte quadam pulcherrimè plicata.* That being opened there was found another cist of lead, apparently in a state of decay, which was thought to be that in which the bones of S. Dunstan, when he was first buried, were inclosed. *Eâ vero apertâ reperta est etiam et alia cista plumbea quasi tabefacta; quæ putatur esse illa, in quâ ossa Sancti Dunstani, cum primum sepeliabatur, recondebantur.* Within these two leaden cists, when they were opened, a certain small thin plate of lead was first found lying on the breast of the body, on which plate this writing was preserved:—

Hic requiescit Sanctus Dunstanus Archiepiscopus,

and this title was written in Roman characters. *Infra has duas cistas plumbeas, cum aperirentur, primo reperta est quædam parva lamina plumbi, jacens supra pectus corporis. In quâ quidem laminâ continebatur hæc scrip-*

overlook and watch the same; and for their accommodation small lofts or chambers, parclosed with open screenwork, were constructed on high, so as to enable

*tura " Hic requiescit Sanctus Dunstanus Archiepiscopus." Et scribitur his titulus Romanis literis.* Then was found a certain stained cloth, very neat and entire, lying under the body of S. Dunstan. *Deinde repertus est pannus quidem tinctus, nitidus valde atque integer, suppositus corpori Sancti Dunstani.* Which being lift up, there appeared that most holy instrument of the Holy Spirit, vested in pontificalibus, then for the most part decayed. *Quo sublevato, apparuit illud sanctissimum organum Spiritus Sancti, indutum pontificalibus tum pro magna parte consumptis.* Afterward appeared the scull, which was handled and kissed as well by the Lord Archbishop as by the Prior and many of the monks. Of which scull, a part divided from the rest the Lord Archbishop delivered to the Prior, to the intent it might be decently adorned, and placed amongst the relics of the church, to be venerated. *Porro apparuit ibidem testa capitis quæ et tangebatur et osculabatur tam a Domino Archiepiscopo quam a Priore ceterisque quam plurimis de conventu monachorum. Cujus quidem testæ partem a reliquo divisam Dominus Archiepiscopus tradidit Priori, ed ratione ut decentur adornaretur, et inter reliquias ecclesiæ venerandum reponeretur.* Lastly they saw other different bones, as well of the arms as of the ribs, and some portion of the flesh of the same our Patron. All which, in very deed, smelt of a most sweet odour. *Denique videbantur et alia ossa diversa tam de brachiis quam de costis, et nonnulla massa de carne ejusdem Patroni nostri. Quæ revera omnia odore redolebant suavissimo.* After this was done, then by command of the Lord Archbishop the above cist was again firmly inclosed both in wood and lead, with as many nails as could be affixed with safety. *Hæc cum peracta fuissent, tunc ad mandatum Domini Archiepiscopi arca superius dicta iterum clausa est firmissime tam opere ligneo quam plumbeo clavisque quam plurimis tutissime affixa.*

I have abbreviated as much as I felt I could the account given by this interesting document. In the meeting of the Archæological Institute at Canterbury in 1844, I carefully noticed the then so-called tomb of Archbishop Theobald, and was persuaded it was a shrine. I had not then met with the document published by Wharton in the *Anglia Sacra.* In the last meeting of the Royal Archæological Institute at Canterbury in 1875, I carefully measured this stone structure. I found it externally to be 7 feet 9 inches long, and 2 feet 10 inches wide. This will fairly agree with the dimensions stated in the document, 7 feet in length, by 1 foot 6 inches in width. The remains of S. Cuthbert in Durham Cathedral were discovered by the Rev. James Raine in 1827. A full and most interesting account of this discovery was published by that gentleman in 1828. The shrine of S. Cuthbert had been defaced on the suppression of the monastery.

the shrines to be overlooked. Of these lofts, that in St. Albans Abbey church, now cathedral, and that erected near the site of the shrine of S. Frideswyde in Oxford Cathedral, are still existing. The feretory or shrine of S. Richard of Chichester, canonized A.D. 1262, was behind the high altar; above the reredos was a watching loft, to which access was obtained from the triforium by three stairs on either side. In Rochester Cathedral is a shrine-like tomb, apparently of the same age as that in Canterbury Cathedral. This has been assigned as the tomb of Bishop Glanville, A.D. 1185—1214, but I think it was probably a shrine, possibly that of S. Paulinus or S. Ithamar.

Eastward of the presbytery we sometimes find the LADY CHAPEL, especially constructed in honour of the Blessed Virgin, but on the Suppression ofttimes converted to secular purposes, as that at St. Albans Abbey, recently used as a grammar school, enriched perhaps with more internal imagery now remaining than in any other church in the kingdom. At Chichester Cathedral the lady chapel is, or was lately, used as a library. Sometimes the lady chapel was not at the extreme east end, but on the side of the choir, as at Ely Cathedral, and at the Collegiate church of S. Mary at Warwick, where the chapel was also a mortuary chapel. There was also a portion of the church used as a SACRISTY or vestry. In some instances this appears to have been between the south transept and the chapter house, and in others to have been behind the high altar. At S. Mary's Abbey church, York, it appears to have been annexed as a distinct room to the south aisle of the choir, and on the south side. At Furness Abbey church the vestiarium or revestry appears to have been a long apartment on the south side of the choir. In some instances the revestry appears to have been at the east end of the conventual church, but so low as not to interfere with or obstruct the great east window. This arrangement appears to have been formerly the case with regard to the Priory churches of both Great and Little Malvern, Worcestershire, though in both cases this building has been demolished. In this position it still exists at the east end of the lady chapel of the collegiate church of S. Mary at Warwick.

The eastern part of the choir, which was ascended by steps, was often vaulted beneath; and the crypt or under-croft generally consists of the most ancient portion of the building.

In different parts of a conventual church, more particularly in the retro-choir or presbytery, were the monuments of noble and munificent benefactors who had chosen that particular church as a resting-place for their remains, and by will had bequeathed their bodies there to be buried. Monuments to Abbots and Priors were also thus placed. In many instances, however, the chapter house was the burial-place of the founder, as well as of many successive heads of the conventual establishment. In the cemetery garth or cloister court were buried the monks, in many instances, at least up to a certain period, without coffins, their bodies, clad in their monastic habits, being simply deposited in the earth, with sometimes a flat or ridged-shaped stone laid over the grave, with a cross of some device or other sculptured in relief or incised thereon, and generally without, but sometimes bearing a short inscription. But in the general devastation and ruin of the conventual churches—those only being wholly or partially excepted which had been or which afterwards became parochial—by far the greater number of these monuments were destroyed, though we now and then meet with broken fragments; and some were removed to neighbouring churches. At Furness Abbey is, or was, the most interesting collection and greatest number of sepulchral memorials I have ever met with at any conventual church in ruins.

MONASTIC COSTUME. Many sepulchral effigies of Abbots and Priors of conventual establishments were destroyed on the suppression of monasteries, and the fragments buried in the ruins. Those remaining are for the most part represented as vested for the altar; and we have but few effigies of this description in which the monastic costume of the particular Order to which such Abbot or Prior belonged takes the place of the vestments for the most important services of the church, and which were common to all Orders alike. The series of effigies of Abbots in Peterborough Cathedral, a fine Benedictine conventual church, represents them severally vested for mass; but at the gate house south-west of that church, over the gateway, is an erect effigy, larger than life, of one in the garb of a Benedictine, and this probably may have been intended to represent S. Benedict himself. Here we have an exact representation of the monastic costume of a Benedictine. The habit consists of a long black robe or cowl, *tunica talaris*, with wide sleeves, and a hood over the head forming part of the cowl. This is apparently of the thirteenth century.

At Orton-on-the-Hill, Leicestershire, removed, I think, on the Suppression from the Abbey church of Merivale, Warwickshire, is the sepulchral effigy of an Abbot of the Cistercian Order, in the habit of that Order. He is represented in the *cappa clausa*, or cowl without sleeves, with the hood thrown back about the neck, and arranged so as to cover the breast and shoulders, whilst on the head is worn the coif or close-fitting scull cap, called the *biretum*. Over the head is an ogee canopy, which refers the date to the fourteenth century. I have only met with one other sepulchral effigy of a Cistercian Abbot, and this lies in the ruins of the Cistercian Abbey of Dundrennan, Kirkcudbrightshire, North Britain, in a fairly perfect state.

In Hexham Priory church, Northumberland, is a well-known sepulchral effigy of a Prior of Canons of the Order of S. Augustine. This has been assumed to be the effigy of one Prior Richard, but it is, I think, of a date long posterior to any Prior of that name. We are told that the habit of this Order, *Vestitus canonicorum est tunica candida cum linea toga sub nigro pallio; tegumentum a scapulis impositum, cervicem totumque contigit caput præterquam a fronte.* With which description, excepting as to colour, for there is none left, this effigy agrees, the hood of the mantle or cloak being drawn over the head.

On excavations being made a few years ago on the site of Dale Abbey, Derbyshire, an abbey of Premon-

Sepulchral Effigy of a Cistercian Monk of Merivale Abbey, now in the Church of Orton-on-the-Hill, Leicestershire.

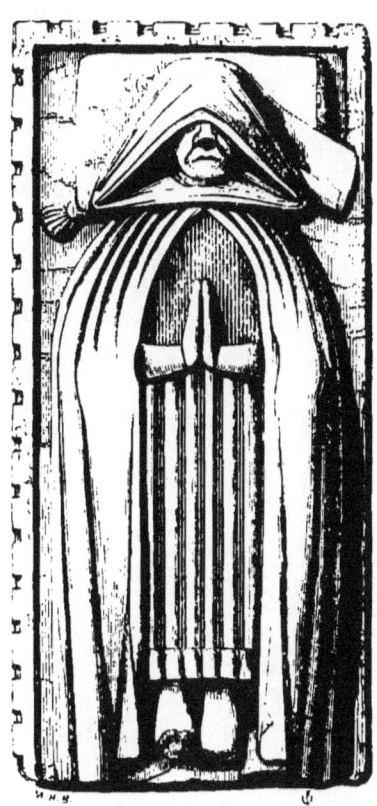

Sepulchral Effigy of Monk of the Canons of the Order of S. Augustine, in Hexham Priory Church, Northumberland. 15th Century.

stratensiars or White Canons, a reformed Order of the rule of S. Austin, a most interesting sepulchral effigy of an Abbot in his monastic garb was discovered within the precincts of the chapter house. He is represented with his head and neck bare, in a long white cassock, *toga talaris candida*, with wide sleeves and numerous folds, but without any hood or mantle, holding in front of the breast with both hands a book. This effigy I take to be of the early part of the fifteenth century.[¶] The white habit of this Order is said to have been ordained by the Blessed Virgin,—dat candidam Maria vestem.

I have not met with any sepulchral effigy of a Franciscan or Dominican Friar, Grey Friars, and Black Friars. Small representations of one of each of these Orders, carved in wood, appear in the subsellia of some stalls in S. Mary's church, Beverly,—that of the Franciscan habited in his cowl and knotted girdle, that of the Dominican in his cowl and scapular in front. In the church of Conington, Huntingdonshire, is a very curious recumbent sepulchral effigy of the fourteenth century, representing a knight who had in after-life taken, in compliance with a well-known custom of that age, the habit of a Franciscan Friar. He appears in a hooded hawberk or tunic of chain mail, the coif and mufflers or gloves of which are alone visible; over this, his defence against human foes, he wears as defensive armour, as it was supposed, against the powers of darkness, the *cappa manicata* or friar's cowl with sleeves,

[¶] Engraved in Vol. i. p. 110 of the *Journal of the Derbyshire Archæological and Natural History Society.*

girt about the waist with a knotted cord, which falls pendant to the feet; over the coif of mail is the *caputium* or hood, the lower portion of which covers the shoulders and upper part of the breast. Below this the hands are conjoined on the breast in attitude of prayer. That the body of the knight thus represented was actually clad in this habit for burial, the hawberk of mail excepted, is more than probable if we investigate the motive why he should be, and that he was one of those in allusion to whom the great poet of the seventeenth century observes—

> "And they who, to be sure of Paradise,
> Dying, put on the weeds of Dominic,
> or in Franciscan thought to pass disguised."

But to refer to earlier authorities, Wadding, in his *Annales Minorum*, tells us that Clement the Fifth, who occupied the papal chair from A.D. 1305 to A.D. 1314, remitted to those buried in the habit of a friar the fourth part of all their sins: *sepeliendis in habitu minorum quartam partem omnium peccatorum remitit.*[r] Wyckliff, who lived in the latter part of the fourteenth century, has the following passage:—"And here men noten many harmes yat Freris doen y the cherch, . . . . but kepyng of Godde's mauntements thei chargen not halfe so muche as he schulde be holden apostata, that lefe ye abite for a daie, but for levyng of dedys of charite, schulde he nothynge be blamed; and thus thei blasfemers in God, and scien whoso dieth in this abyte schall never go to helle for holyness that is therein." In a letter from Latimer, Bishop of Worcester, to Sir Edward

---

[r] *Annales Minorum*, Vol. iii. p. 148, A.D. 1313.

Sepulchral Effigy of a Knight in a Friar's Habit, in Conington Church, Huntingdonshire.

Baynton, is the following passage:—"I have thought in time past that if I had been a friar and in a cowl, I could not have been damned nor afraid of death." And in a sermon preached by him in 1537, he says, "But yet they that begot and brought forth our old ancient purgatory pickpurse; that was swaged and cooled with a Franciscan's Cowl; put upon a dead man's back to the fourth part of his sins."

And elsewhere in a sermon preached by him on Septuagesima Sunday, in 1552, he observes as follows:— "True it is that God requireth good works of us, and commandeth us to avoid all wickedness. But for all that, we may not do our good works to the end to get heaven withal; but rather to show ourselves thankful for that which Christ hath done for us, who with His passion hath opened heaven to all believers; that is to all those who put their hope and trust not in their deeds, but in His death and passion, and study to live well and godly; and yet not to make merits of their own works, as though they should have everlasting life for them, as our monks and friars, and all our religious persons were wont to do, and therefore may rightly be called murmurers; for they had so great a store of merits that they sold several of them unto other men, and many men spent a great part of their substance to buy their merits, and to be a brother of their houses; or to obtain one of their coats or cowls to be buried in." In the *Defense of the Apologie of the Church of England*, published in 1570, against *The Confutation and Detection of Sundry Foule Errors, &c.*, written by that great controversialist and opponent to Bishop Jewell, Dr. Harding, who denied that the Catholics—that is those of the Roman Church—put great holiness in mere outward observances, as in apparel, in which defence Jewell puts this question to Harding, "Wherfore doothe Thomas of Aquine tel us so certainly that the wearinge of Francise or Dominike's Cowle had power to remove sinne as well as the Sacrament of Baptisme?"

The marginal reference given by Jewell to the works of that great schoolman is a general one, and I have been unable to verify it.

Amongst the works of Becon against the Church of Rome, those on *The Acts of Christ and of Anti-Christ*, published in 1577, contain the following passage:— "Anti-Christ for the forgiveness of our sins and of our justification, sendeth us to his pardons and Bulls, to his years of Jubilee and masses of *Scala Cœli*; yea, he sendeth us to a Grey Friar's Cowl and willeth us to be buried in that, promising us by that means both remission of sins and everlasting life."

Other extracts might be given shewing the belief in and prevalence of the practice.

The only other effigy of a friar I am at present conversant with is a small demi-figure acting as a crest to a tilting helme beneath the head of a knight in Sawtry All Saints' church, Huntingdonshire, a church in the neighbourhood of Conington. This figure constitutes a portion of an incised brass, and is represented habited in the *cappa manicata* or sleeved cowl, with the *caputium* or hood attached and drawn over the head; in the hands the ancient "discipline," as the instrument in the shape of a whip with knotted cords was called, is represented as held. This figure is of the early part of the fifteenth century, the date given on the inscription being 1404.

Of NUNS in their peculiar apparel we have but few sepulchral effigies, either sculptured in stone or incised in brass. In the church of Polesworth, Warwickshire, the conventual church of a Benedictine Nunnery, is the rare sculptured sepulchral effigy of an Abbess of that Order. This is, I think, of the thirteenth century. Her head appears in a trefoil-shaped sinking. Round the head, chin and neck, is the coif and wimple, and over the head is worn the veil. She is habited in a large gown or cowl, but without any mantle. With her right hand is grasped the pastoral staff, floriated within the crook, which is turned inwards; and in her left hand is held the *horæ* or book of prayers. Beneath her feet the figure of a hart or stag is represented, probably in allusion to a verse in the Psalms—" As the hart desireth the water brooks," &c. In the church of Elstow, Bedfordshire, which was like that at Polesworth, the conventual church of a Benedictine Nunnery, is the sepulchral incised brass of Elizabeth Hervey, Abbess, who died about the year 1524. The incised effigy represents her clad in a cowl or large loose gown with wide falling sleeves, beneath or within which appear the close-fitting sleeves of the inner vest. Over the gown is worn the mantle, but how this is fastened is not apparent. Over the chin and in front of the neck is worn the plaited wimple, not like the ancient wimple or neck and chin cloth, but like the widow's barbe or chin cloth of the latter part of the fifteenth or early part of the sixteenth century. On the head is a coif and veil, the latter falling down on each side to the shoulders. The hands are uplifted and conjoined together in prayer. The pastoral staff, with the curved crook, the latter somewhat richly designed, turned outwards, rests within the right arm and elbow. The plaits of the wimple, or neck and chin cloth, remind us of the Prioress described by Chaucer, in the fourteenth century, "*Ful semely hire wimple ypinched was.*"

On the sides of the rich tomb in Oxford Cathedral, on which lies the recumbent effigy of Elizabeth Lady Montacute, who died A.D. 1353, are small statuettes, eighteen inches in height, representing her children. Two of her daughters were, in succession, Abbesses of the Benedictine Nunnery of Barking in Essex. One of these is represented (for these statuettes were coloured) in the white cowl, a long loose gown or robe, with a black mantle over, connected or fastened in front of the breast by a chain. The head of this statuette has been destroyed, but remains of the plaited, or, to use Chaucer's words, "*ypinched*" wimple—an early instance, I think, of the wimple being plaited, in accordance with Chaucer's description—which covered the neck and chin, are visible, as are also portions of the white veil falling on the shoulders. The pastoral staff appears on the left side, within the left arm, but the crook is gone. The effigy of the other daughter, also an Abbess, is in most respects similar to the former, with the exception that the left sleeve of the cowl or gown—which sleeve is large and wide—is seen as well as the close-fitting sleeve of the inner vest; but in the former statuette the close-fitting sleeve alone appears.

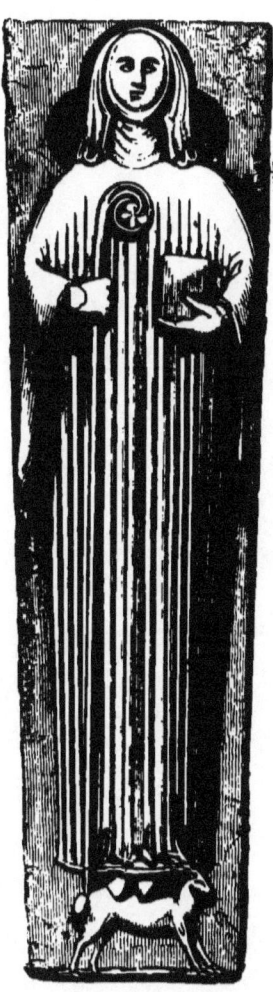

Sepulchral Effigy of an Abbess of Polesworth Nunnery
in Polesworth Church, Warwickshire.

In that splendid illuminated folio, the well-known *Louterell Psalter*, supposed to have been executed in the early part of the fourteenth century, amongst other representations is that of an Abbess holding in her right hand the pastoral staff with the crook turned outwards, her left hand being upheld, with three fingers upraised in act of benediction. Her cowl or gown and wimple are white, but the wimple is not plaited, or "*ypinched*," her mantle is black, and on her head is a veil falling down on each side to the shoulders.

In one of the Arundel MSS. in the British Museum, the habit of the Sisterhood of Syon Monastery, an establishment of Nuns of the Augustine Order of S. Bridget, is described in relating the manner in which the body of each professed inmate of that house was, on her decease, to be prepared for burial. And although this house was not, like those at Polesworth and Elstow, of the Benedictine rule, this MS. throws some light on the habit of both Orders. It states as follows:—" They schal clothe the body with cowle, and mantel, veyle and crown, without rewle cote, but with hosen and schone tanned, and with a gyrdel, whiche al schal be of the vileste gere, and in al these, except mantel, sche shal be buryed."

Such is a summary of the particular apparel, both male and female, of the principal monastic Orders, as displayed by sepulchral effigies, which are indeed the most trustworthy sources on which we can rely for an elucidation of ancient monastic costume.

Even before the general suppression many of the smaller conventual establishments had been unsatisfactorily governed, and plate and goods had been made away with by those who presided over the same. As to the alien priories these had been subject, even in the fourteenth century, to confiscation in time of war between this country and the country abroad to some of the larger monasteries of which they had been affiliated. Some of the larger and most richly endowed monasteries, as those of Glastonbury, St. Albans, and Durham, kept up their state to the very last. Some nunneries, as at Polesworth and Catesby, were most favourably reported of by the Commissioners of Inquiry; but the good reports of such did not avert the general ruin to which they were destined.

The inventories of goods, taken at the Suppression, especially of the minor houses, and even of some of the large Cistercian monasteries, shew them to have been poorly furnished, as those of Pipewell Abbey, Northamptonshire, and Merivale Abbey, Warwickshire; whilst at Sele Priory (Beeding), Sussex, hardly any goods were left at all, it having been deserted by its last occupants, Carmelites.

Many of the religious houses were greatly in debt, and valuable articles of plate and otherwise belonging to them had been disposed of previous to the Suppression. Though the ostensible break between the Pope and the monarch of the crown of England may have been immediately owing to personal causes, there can be no doubt but that these religious institutions were gradually decreasing in public estimation, and that ere long very stringent reforms, to say the least, would have been needed. The confiscation by Wolsey, before his downfall, of certain religious establishments, the revenues of which were applied towards the endowment of the new colleges founded by him at Ipswich and Oxford, were but the precursors of the general suppression of the monasteries, first in 1537 of the lesser, (that was of those the revenues of which severally were under £200. per annum), and then in 1539 of the greater monasteries.

Of the goods belonging to some of these religious houses, the following inventories, made on their suppression, may give an idea of their substance, the farming stock not being included.

"The White fryers (Carmelites) of Sele, (Beeding Sussex), A.D. 1537. The Inventory of Sele the xvi day of July, priceyed by Sir Raffe Berneys, Vicar ther, Henry Blurners, Edmund Grene, John Grenear, and John Erlye.

"Alle the stuff ther ys a spete, a sory belle, iii or iiii old formeys, ii or iii raggd cheseabulles and tenakylles, all priceyd at iii$^s$ iiij$^d$. Ther ys a lytell belle in the parysche stepulle, the whyche the freeres useyd but the parysche sathe that yt longyth to them, but yt ys priceyd at vi$^s$ viii$^d$. Ther be stalles in the quere worth xx$^s$. Thys ys alle the holle substans of that house."

Ornaments in the church and vestry of the Abbey of Pipewell, Northamptonshire, at the time of the suppression of that abbey in 1539.

"Fyrst upon the hygh alter, 1 table of Alebaster, 1 table of carvyd tymbr w$^{th}$ great images and in yt 3 great candlestykes of latenn, 2 deskes of latenn, the munkes stales in the same quere, 1 payr of orgaynes,

3 litell deskes in Seint Benettes chappell, 1 oulde table of woode, 1 chest, 1 sete ther, 1 pticōn of tymbr̄, 1 table of alebaster in Seinte Stephens chapell, 1 pticōn of tymbr̄ ther, 1 imag' of oͬ Lady of Pytye, 2 tables paynted in Seint Michelles chapell, 1 table of alebaster, 1 image of King Henry. In Seint Nicholas chapell 1 table of wood payented, and 1 image of Seint Nicholas. At the Trynytye alter, 1 table of alebaster, 1 image of the Trynytye. At Seint Katheryns alter, 1 table of alebaster, certain ould images, 2 lytell ieron candlestykes in the walle. At Seint Peters alter, 1 table of alebaster, 1 lytell image of oͬ Lady, 1 pticōn of tymbr̄. At Seint Mary Magdelyn's alter, 1 table of wood payented 1 pue, and the pticion of tymbr̄. In the body of the churche, 1 rode and the pticions of 2 lytell ould holy walter stokes of lead, 1 of brasse, certain stoles and ould cupbordes in the upp ende of the churche. In the belfry 1 cloke and the cloke howse. In the dorter, the munkes selles and 1 laumpe of laten. In the vestry the vestments are enumerated. The whole sould for £16. 3s. 0d."

The different offices are thus mentioned: "The cloyster with the munkes seates; the frater—this appears to have been the refectory, as it contained '1 pulpytt;' the haule; the perlore; the buttery and pantry; the kychynne; the larder; the fyshehouse; the bruehouse; the gelynghouse; the maultehouse; the bakehouse and bultynghouse; the smythes forge; the hall chambr̄; the next chambr; the nether churche chambr; the hygh churche chambr; the newe chambr; the stody chambr̄; the sᵉ vātes chambr." The establishment, including the Abbot, consisted of fourteen monks and forty-three servants. The chapter house is not mentioned. Of gilt plate belonging to the church there was 70 oz., and of whyte plate, 271 oz.

The furniture of the church of the Cistercian Abbey of Merivale, at the time of the Suppression, is described by the King's Commissioners as follows:—

"Fyrst a table of allablaster.

"It. ij candelstyks of latten.

"It. onne laumpe of laten.

"It. the monks seats of tymbr.

"It. a payre of organnys.

"It. vi olde alters with imagis.

"It. the pticion of olde tymber in the body of the curche.

"It. thre iron candelstyks before the alters.

"It. a holy water stoke of brasse.

"It. the glasse and the iron in the wyndoys of the curch.

"It. all the pavent in the curche.

"It. vj. grave stones wyth brass in them."

These were sold for £4. 11s. 8d. The vestments in the westrie sold for £6. 9s. 0d.

Then are mentioned the cloyster and the chapt house.

"It. xxviij panys of paynted glasse 0 : 5 : 0.

"It. a laver of ley mettall and leade before the same laver.

"The hall; the buttery; the cheffe plor; the inner chamber; the greate old chamber; the chamber next the old chab; the chamber callyd ye Brcdonnes; the whyte chamber; the porter's chamber; the kechyn; the larder house; the brewhouse; the maltehouse; the bakehouse; the lymehouse. Gylte plate, 132 oz., whyte plate, 73 oz." 4 bellys were valued at £30. 0s. 0d. Ten monks, including the Abbot, were pensioned off.

Having given the inventory of goods in the two Cistercian Abbey churches of Pipewell and Stoneleigh at the time of the Suppression, I proceed to add the contents of a grand Benedictine church, that of Peterborough, now a cathedral, as taken in 1539.

"IN THE CHOIR. Imprimis, the high altar, plated with silver, well gilt, with one image of Christ's Passion, and a little shrine of copper, enamell'd for the sacrament.

"Item, two pair of organs, and two desks of latten, seven basons hanging, with four candlesticks, and banners of silk above the choir, joining to the tomb where Queen Katherine lies buried.

"Item, in the inclosed place, where the Lady Katherine lieth buried, one altar-cloth of black cloth, one pall of black velvet, with white cloth of silver crossed, and one white altar cloth.

"Item belonging to the same remaining in the sacristry, two candlesticks of silver, parcel gilt, one chalice and two cruets gilt.

"Item, one pair of vestments of black velvet with an alb to the same.

"Item, ten cloths called Pede cloths to lie before the high altar.

"Item, at the upper end of the church three altars, and upon every altar a table of the Passion of Christ, gilt, with three stained fronts.

"IN THE LADIES CHAPPEL. Imprimis, an image of our Lady with reddis rissey, set in a tabernacle well gilt, upon wood, with twelve great images, and four-

and-thirty small images of the same work about the chappel.

"Item, a pair of organs, one desk and four seats, one Tabernacle of the Trinity, and one other of our Lady, one desk, and one old candlestick of latten, four pede-cloths called tapets.

"Item, two vestments of white damask with flowers, one red vestment of satin with flowers and also albes for the same.

"Item, one suit of crimson velvet, with orphers of imagery of gold, and one cope and four albes.

"Item, three white altar cloths, one of them diaper, with three old painted fronts, two orphers, eight surplices.

"In S. John's Chappel. Item, a table of alabaster, one front of painted cloths, with two images of alabaster.

"In S. James' Chappel. Item, one table of alabaster, two images of the same, and one front of painted cloths.

"In the Rood Loft. Item, one table upon the altar, eighteen images well gilt, one desk of wood, two orphers, one front of painted cloth.

"In the Body of the Church. Item, one altar with images, gilt, one front of green silk, with ostrich feathers, one coffer, the altar of our Lady's Lamentation, gilt, one front of painted cloths, four lamps in divers parts of the church.

"In the South Ile. Item, in S. Oswald's chappel, one altar with a front of painted cloth, one table gilt of S. Oswald.

"Item, in S. Bennet's chappel, one altar with a front

of painted cloth, one table gilt with the story of S. Bennet.

"Item, in S. Kyneburgh's chappel, one altar with a front of painted cloth, with one table well gilt.

"THE TRINITY CHAPPEL. Item, the altar with a front of old silk, one white altar cloth of diaper, two candlesticks of latten, one table of alabaster, one coffer, and seats of wood, one lamp.

"IN THE CLOYSTER. Item, one conduit, or lavatory of tynne, with divers coffers and seats there.

"IN THE OSTRIE CHAPPEL. Item, one altar cloth, two fronts of painted cloths, two latten candlesticks, one coffer, one super-altar of marble, one vestment of green silk, one vestment of dove-colour'd silk, two albes.

"IN THE CHAPPEL OF LOW. Item, one altar, two fronts of old painted cloths, two vestments of silk, one red the other green, and one albe of needle-work.

"Item, one chalice gilt, one corporas, one pair of candlesticks of latten, three bells to ring in the chappel.

"IN THE INFIRMARY CHAPPEL. Item, one table of alabaster, one front of silk, two candlesticks of latten, three coffers, four seats, one vestment, one albe of white silk, with orfers of red.

"Item, one vestment of white fustian for Lent, with an albe, three corporasses with cases, two altar cloths, one old vestment, one old albe.

"Item, one little bell, one lamp hanging, one broken silver crown.

"Item, old cloths to cover saints in Lent.

"IN THE ABBOT'S GALLERIE CHAPPEL. Item, one table of alabaster.

"In the other Chappel. Item, one table of alabaster, three old chests, an old almery."

The plate, consisting of crosses, candlesticks, censers, ship, chalices and patens, and the vestments, as enumerated, are too numerous to specify severally.

Amongst the buildings are enumerated, with their dimensions: the church; Lady chappel; other chappels; cloister; the great dormitory; the little dorter; the fratry; the infirmary; the chappel at the gate of the monastery; the vestry; the Abbot's hall; the Abbot's great chamber; the Convent's kitchin.

The tables of alabaster so frequently adverted to were sculptured in bas-relief, serving as the reredos to the various altars.

Many of the existing remains of our ancient monastic piles, some from recent excavations, have of late been planned out, and it is from the plan of one of these, the Cistercian Abbey of Kirkstall, near Leeds, that we may be the better able to understand the general arrangement of a Cistercian Abbey, with its offices.

Thus I have endeavoured to give a succinct, though imperfect and crude notion of ancient monastic arrangement in this country, any general rules respecting which must be modified by numerous exceptions. Though we now see our ancient monastic remains,—those which are not incorporated with mansion houses, which is the case with many,—in a state of ruin, exposed to and beaten down with the storms of three centuries and a half, and the violence of man, they are now, as a rule, better cared for, and in not a few cases their sites have been excavated and preserved.

They appear to me to be of more interest than many of the monastic remains in France, which were devastated in the Revolution at the close of the last century. For the monastic offices in many of the French abbeys were re-built in the Palladian style, apparently some time in the latter part of the sixteenth or early part of the seventeenth century, after the fashion then prevalent. Possibly in consequence of the destruction occasioned by the Huguenots in the sixteenth century; and which, by occasioning a revolution of feeling, barred their efforts for church reformation. The comparatively modern buildings, however more commodious than those in the room of which they were erected, do not excite in us the same imaginative feelings the older buildings of our own country are apt to do. In the latter we find evidence demonstrative of the peculiar manners and customs of those religious communities, bound together by vows and rules, which exercised so much influence during the Middle Ages; whose faults are inveighed against in the satirical writings of the medieval poets, and noticed in the monastic records themselves, the latter furnishing at least unbiased, if not unconscious, evidence against them. On the other hand, if we regard them as the foundations of pious men, in ages, compared with our own, rude and barbarous; as the chief instruments in those ages of civilization; for diffusing religious and secular knowledge; for the encouragement of literature; for the exercise of hospitality; as adepts in painting and sculpture; as agriculturists; or as chroniclers of their own and by-gone times, we do them but simple justice. But we

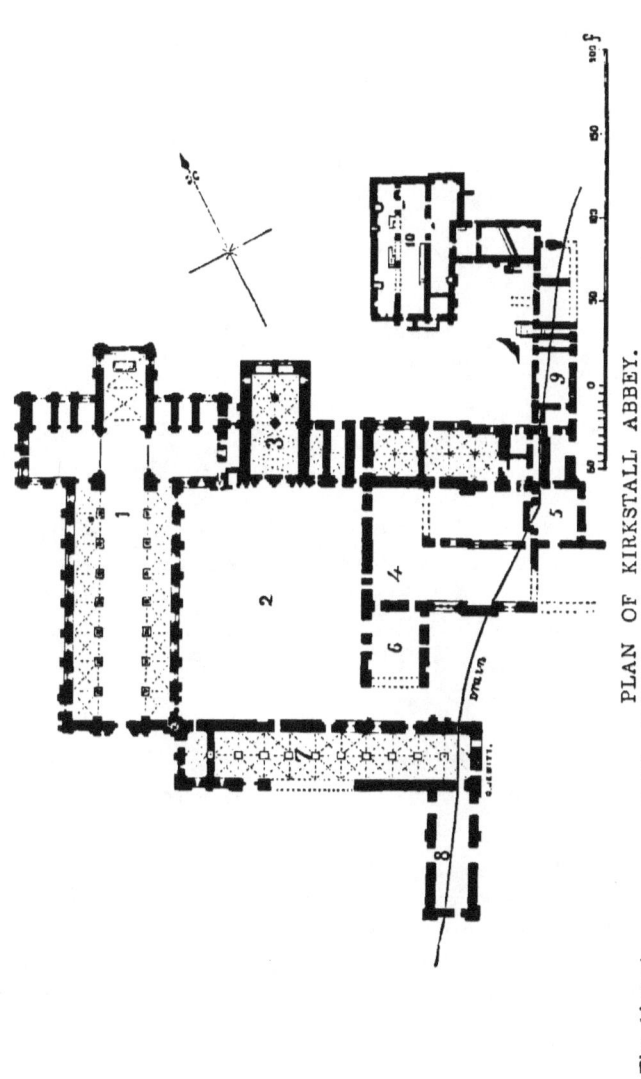

PLAN OF KIRKSTALL ABBEY.

1. The Church.
2. The Cloister Court.
3. The Chapter House.
4. The Refectory.
5. The Kitchen.
6. The Locutorium, or Common Room.
7. The Ambulatory under the Dormitory.
8. The Supposed Infirmary.
9. The Abbot's Lodgings.
10. The Hospitium, or Guest House.

live in far different times; and whilst we may derive many an important lesson from the past, and agree with the Christian philosopher of Norwich, that "It is opportune to look back upon old times, and let nothing remarkable escape us," the present is hardly an age which requires continuous monastic seclusion, but rather of study, combined with active exertion, in the duties of our several callings; and demands of us each in his station the verification of that other aphorism, " Up and be doing, and God will prosper."

Sepulchral Headstone Cross, Goodnestone, Kent.

# GENERAL INDEX

## TO CHURCH ARRANGEMENT, FURNITURE, AND MATTERS.

## VOL. II.

ALTARS, 10. 139.
Altars, High, 76.
Altars, stone, 77.
Altars, form of, 77.
Altar lights, 51. 80.
Ambrie, 96.
Ampulla, 85.
Anglo-Saxon Churches, 8.
Aspersorium, 10.

BANGOR, USE OF, 6.
Bells, 23.
Bells, sacring, 26.
Bells, benediction of, 28.
Benedictine Order, 235.
Bestiaries, 46.
British Church, ancient, 8.

CANONS of the Order of St. Augustine, 231.
Campanology, 29.
Capellæ Carnariæ, Charnel Vaults, 185.
Carmelites, 236.
Carthusians, 236.
Chalice, 86.
Church furniture, 49.
Church Chests, 157.
Chrystmatory, 85.
Chantries, and Chantry Altars, 131.
Cistercian Order, 235.
Confessionals, 124.
Constitutions Apostolical, 4.
Credence, 96.
Crewets, 83.
Crosses, Dedication, 155.

DESKS, 71.
——— Brass, 71.
——— Eagle, 71.
——— Stone, 75.
——— Wooden, 71.
Dominican Order, 235.
Domus Inclusi, 163.

EARTHEN JARS, 154.
Easter Sepulchres, 98.
Encaustic tiles, 228.
Encaustic tile Kilns, 232.

FONTS, 16.
Fonts of lead, 17.
Font covers, 20.
Franciscan Order, 236.

GALLICAN LITURGY, 5.
Gelasian Liturgy, 5.
Gilbertines, 236.
Gregorian Liturgy, 5.

HAGIOSCOPE, 147.
Hearts enshrined, 148.
Hereford, Use of, 6.

IMAGES, 54.
Images, introduction of, 55.
Image brackets, 65.

KNIGHTS TEMPLARS, 236.
Knights Hospitallers, 236.

LADY OF PITY, 59.
Leonine Sacramentary, 4.
Lincoln, Use of, 6.
Liturgical Offices, early, 2.
Low side Windows, 127.

MONASTIC ARRANGEMENT—
　The Close, 246.
　The Gatehouse, 246.
　The Chapel of the Gatehouse, 248.
　The Dormitory, 250.
　The Refectory, 253.
　The Refectory Pulpit, 254.
　The Cellerage, 255.
　The Lavatory, 256.
　The Kitchen, 258.
　The Abbot's or Prior's Lodgings, 258.

# INDEX TO CHURCH ARRANGEMENT.

MONASTIC ARRANGEMENT—
  The Chapter House, 262.
  The Mortuary Chamber, 264.
  The Cloisters, 266.
  The Locutorium, 268.
  The Scriptorium, 268.
  Carrels, 268.
  The Library, 268.
  The Treasure House, 270.
  The Infirmary, 270.
  The Hospitium, 270.
  The Bakehouse, and Brewhouse, 272.
  The Abbey Mill, 272.
  The Sewers, 272.
  The Conventual church, 273.
  The High Altar, 276.
  Shrines, 277.
  The Lady Chapel, 281.
  The Monastic Costume, 283.
  The Monasteries, Suppression of, 300.
  The Goods of Monasteries, Inventories of, 300.
  The Plans of Monasteries, 307.

MISERICORDES, 45.
Monstrance, 83. 85.
Mural paintings, 196.

ORNAMENTS OF CHURCHES, 49.

PAINTINGS IN CHURCHES, 9.
Painted tables, 213.
Painted glass, 214.
Paintings on Rood-loft screens, 207.
Paten, 86.
Pax table, 84.
Peeping Tom of Coventry, 61.

Phœnicia, Christian structures of, 7.
Piscina, 92.
Piscina double, 95.
Piscina triple, 95.
Piscina ground, 96.
Pix, 83. 86.
Porches, 10. 14.
Premonstratensian Order, 236.
Pulpits, 32.
Purgatory, 78. 132. 133.

RECLUSIO ANACHORITARUM, 168.
Relics in altars, 146.
Reliquaries, 148.
Rood lofts, 38.
Rood loft images, 42.
Rood loft altars, 140.

SARUM, USE OF, 6.
Screen work, 85.
Sculpture, Norman, 66. 67.
Sculptured tables, 68.
Seats in Churches, 29.
Sedilia, 90.
Service books, 50.
Ship, 85.
Stalls, 44.
Stone offertory box, 63. 160.
Stoups, 10.
Subsellia, 45.
Symbols of Saints, 207.

TOWERS, 20.
Thurible or Censer, 85.

VESTIARIUM, or VESTRY, 156.

YORK, USE OF, 6.

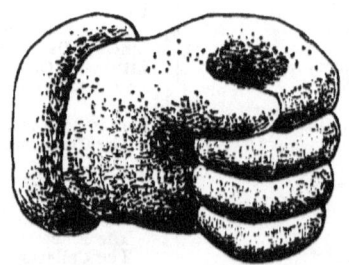

Stone Candlestick, Upton Castle Chapel, Pembrokeshire.
Page 65 ante.

# INDEX

## TO CHURCHES, CATHEDRALS, AND MONASTERIES, ARRANGED IN COUNTIES.

## VOL. II.

### BEDFORDSHIRE.
Bradwell, 21.
Elstow, 21. 295.
Keysoe, 18.
Leighton Buzzard, 71.
Luton, 92.
Marston, 148.
Marston Morteyne, 21.
Stotfold, 202.
Turvey, 92.
Woburn, 21.

### BERKSHIRE.
Ashbury, 84.
Avington, 91. 93.
Childrey, 17.
Uffington, 156. 181.
Windsor, St. George's chapel, 139.

### BUCKINGHAMSHIRE.
Chesham Bois, 219.
Chetwode, 216.
Wing, 8.

### CAMBRIDGESHIRE.
Cambridge, St. Mary's, 69. 125.
Ely priory, 86. 274. 281.
King's College chapel, 226.
Thorney abbey, 238.

### CHESHIRE.
Chester, St. Werburgh's abbey, 243. 246. 250. 251. 253. 254. 255. 258. 268.
Gawsworth, 201. 202.
Nantwich, 181.
Pulford, 86.

### CORNWALL.
Mylor, 21.
Parranforth, 29.
St. Neots, 222.

### CUMBERLAND.
Bridekirk, 18.
Burgh-on-the-Sands, 22.
Carlisle cathedral, 213. 246. 255.
Great Salkeld, 22.
Lannercost priory, 246. 253. 255.
Newton Arlosh, 22.
Wetherall priory, 246.

### DERBYSHIRE.
Ashbourne, 147.
Belper, 144.
Bolsover, 44. 68.
Breadsall, 59. 68.
Dale abbey, 222. 284.
Morley, 222.
Norbury, 219. 220.
Repton priory, 8. 239.
Yolgrave, 20.

### DEVONSHIRE.
Chittlehampton, 21.
Collumpton, 43.
Exeter cathedral, 47.
Holcombe Burnell, 120. 122.
Marldon, 182. 192.
Paignton, 156. 183.
St. Mary Ottery, 71.
Totness, 41.

### DORSETSHIRE.
Bradford Abbas, 137. 161.
Cerne Abbas, 12.
Fordington, 12.
Melcombe Horsey, 200. 202.
Milton abbey, 250.
Pulham, 131.
Sherborne, 44.
Wareham, 17.

### DURHAM.
Brancepeth, 161.

Durham abbey church, 13. 63. 107.
108. 109. 131. 132. 253. 256. 257.
268. 270. 277.
Finchale priory, 159. 253. 255. 256.
262.

### ESSEX.
Barking abbey, 247. 248.
Brightlingsea, 156.
Colchester, St. John's abbey, 247.
270.
Copford, 199.
Eastbourne, 142.
Ipswich, St. Nicholas, 155.
Layer Marney, 138.

### GLOUCESTERSHIRE.
Berkeley, 21.
Cuberley, 69. 150.
Daglingworth, 178.
Deerhurst, 239.
Fairford, 224. 225.
Forehampton, 77.
Gloucester cathedral, 47. 92. 132.
139. 160. 257. 268. 270. 274.
Kempley, 199.
Sedgeberrow, 13. 92.
Tewkesbury abbey, 139. 274.
Thornbury, 19.
Westbury-on-Severn, 21.

### HAMPSHIRE.
Beaulieu abbey, 33. 254.
Brixton, Isle of Wight, 145. 162.
Christchurch priory, 139.
East Meon, 200.
Middleton, 75.
Romsey abbey, 29. 44. 93. 213. 274.
Shanklin, Isle of Wight, 162.
Shorwell, Isle of Wight, 200.
St. Cross, 96.
Winchester, St. John's, 37. 200.
———— cathedral, 45. 75. 139. 276.

### HEREFORDSHIRE.
Burg, 141.
Colwall, 234.
Hereford cathedral, 62.
Kilpeck, 93.
Ledbury, 21.
Peterchurch, 140.
Urinshaw, 140.
Wigmore, 141.

### HERTFORDSHIRE.
Aldbury, 74. 137.
Saccomb, 88.

St. Albans abbey, 90. 142. 199. 201.
246. 265. 274. 277. 281.
300.
———— St. Peter's, 71.
Watford, 162.

### HUNTINGDONSHIRE.
Bury, 74.
Conington, 289.
Eynesbury, 21.
Farcett, 156.
Haddon, 30.
Morborne, 201.
Orton Longueville, 201.
Ramsey abbey, 74. 84.
Sawtry, All Saints', 294.
Standground, 148.
Water Newton, 30.
Yaxley, 149.

### KENT.
Bapchild, 93.
Bobbing, 66. 67.
Boxley abbey, 239. 246. 272.
Brookland, 17.
Canterbury cathedral, 39. 140. 196.
213. 215. 217. 274.
———— St. Augustine's, 86.
———— St. Dunstan's, 12.
Dartford, 183. 202.
Detling, 74.
Doddington, 130.
Dover, St. Margaret's, 129.
———— St. Martin's nunnery, 270.
Farningham, 18.
Folkestone, 195.
Hythe, 194. 195.
Leeds, 37.
Lenham, 145.
Leybourne, 150.
Maidstone, 145.
Malling abbey, 246. 248. 267. 271.
273.
Norfleet, 37.
Rochester cathedral, 202. 281.
Sandwich, St. Clement's, 145.
———— St. Peter's, 192.
Southfleet, 16.
Swanscombe, 75.
Tunbridge, 39.
Westwell, 220.
Wrotham, 182.

### LANCASHIRE.
Claughton, 25.
Croston, 141.

## MONASTERIES REFERRED TO.

Furness abbey, 246. 248. 265. 271. 282.
Preston, 141.
Standyche, 141.

### LEICESTERSHIRE.
Aston Flamville, 26.
Bottesford, 27.
Brentingby, 26.
Buckminster, 182.
Dunton Bassett, 30.
Edmundthorpe, 192.
Evington, 64.
Frolesworth, 131.
Garthorpe, 119.
Glen Magna, 25.
Leicester, All Saints', 26.
——— St. Mary's, 91. 178.
——— St. Martin's, 160.
——— St. Margaret's, 160. 183.
——— St. Peter's, 173. 178.
Lutterworth, 34.
Melton Mowbray, 131.
Orton-on-the-Hill, 284. 285.
Sysonby, 25.
Theddingworth, 88.
Thurcaston, 36.
Twycross, 227.
Ulverscroft priory, 252. 255. 262. 273.
Willoughby Waterless, 26.

### LINCOLNSHIRE.
Asbye, 115.
Baston, 115.
Belton, 115.
Birton, 116.
Blyton, 116.
Bourne, 18.
Bradley, 18.
Castlebyth, 116.
Coates, 42.
Cockerington, 156.
Croxton, 116.
Denton, 116.
Depyng priory, 269.
Durrington, 116.
Fleet, 21.
Flixborough, 21.
Fosdyke, 20.
Giles, St., Hospital of, 58.
Grantham, 131. 181. 191.
Haconby, 161.
Heckington, 120. 122.
Howell, 145.
Keelby, 93.
Kirby, 122.
Lincoln cathedral, 58. 84. 120. 121. 122. 257. 263.
Louth, 162.
Navenby, 120.
Ranceby, 145.
Saltfleetby, St. Peter's, 162.
Somercotes, 25.
Stow, 42.
Tapholme priory, 254.
Thornton abbey, 246.
Uttoft, 161.
Wellingore, 91.

### MIDDLESEX.
Hadley, 23.
London, St. Paul's cathedral, 188.
——— White Friars' priory, 269.
——— Grey Friars, 55.
——— St.George the Martyr, 93.
——— St. Michael's, Cornhill, 39. 65. 125.
——— St. Peter's, Tower of London, 200.
Stanwell, 105.
Westminster abbey, 48. 131. 264. 277.

### NORFOLK.
Batfield, 69.
Bradeston, 182.
Castle Acre priory, 243. 259.
Cromer, 181.
Deopham, 16.
Ditchingham, 202.
Fakenham, 18.
Filby, 16.
Frenze, 185.
Gessing, 185.
Great Yarmouth, 137.
Hingham, 16.
Ingham, 204.
Martham, 16, 181.
Morley, St. Botolph, 16.
Norwich cathedral, 14. 257. 267. 274.
——— St. Edward's, 170.
——— St. Etheldred's, 170.
——— St. John Sepulchre, 18.
——— St. John the Evangelist's, 171.
——— St.John the Baptist's, 171.
——— St. Julian's, 170.
——— St. Gregory's, 191. 202.
——— St. Peter's, 69. 154.
North Walsham, 20.
North Wold, 120. 122.
Outwell, 181.
Raineham, 105. 106.
Ranworth, 182.
Rickenhall, 16.

South Burlingham, 75.
South Creak, 16.
Terrington, St. Clement's, 21.
Thorp Abbots, 182.
Trunch, 16. 20.
Tunstead, 16.
Walton, 21.
Walpole, St. Peter's, 181.
Westwick, 16.
Wigenhall, 16. 145. 181.
Westwick, 16.
Worstead, 20. 181.
Wymondham abbey, 238.

### NORTHAMPTONSHIRE.

Ashby St. Ledgers, 20.
Aston, 97.
Barneck, 65.
Brixworth, 8. 151. 152.
Byfield, 30.
Catesby priory, 30.
Clopton, 25. 26.
Cold Ashby, 25.
Cold Higham, 21.
Dallington, 128.
Denford, 154.
Dodford, 204. 205.
Earl's Barton, 91. 97.
Ecton, 12.
Elton, 156.
Fawesly, 26.
Finedon, 30.
Floore, 97.
Geddington, 37.
Higham Ferrers, 230. 275.
Irthlingborough, 12. 148.
Kilsby, 116.
Little Billing, 18.
Loddington, 25.
Lowick, 220.
Maidford, 21.
Maxey, 141.
Norborough, 192.
Oundle, 71.
Peterborough, 246. 256. 266. 283. 304.
Pipewell abbey, 61. 248. 253. 255. 300. 301.
Rothwell, 92. 95. 194. 195.
Spratton, 92.
Stanford, 220.
Strixton, 156.
Thorp, 27.
Thrapston, 26.
Ufford, 30.
Walmsford, 17.
Whitwell, 145.
Wittering, 8.

Woodford, 151.
Wood Newton, 26.
Yarwell, 30.

### NORTHUMBERLAND.

Hexham priory, 8, 9. 239. 284. 286.
Jarrow, 8. 10. 239. 243.
Monkswearmouth, 8. 9. 14. 214. 239.

### NOTTINGHAMSHIRE.

Claypole, 143.
Cortlingstock, 129.
Hawton, 31. 120. 121. 122.
Holme, 184.
Newstead, 72.
Sibthorp, 120.
South Collingham, 145.
Southwell, 72. 92.

### OXFORDSHIRE.

Adderbury, 132. 144.
Adwell, 150.
Bampton, 161.
Beckley, 148.
Begbrook, 21.
Bloxham, 91.
Burford, 44. 201.
Chipping Norton, 148. 180.
Cropredy, 37. 71. 183.
Dorchester, 17. 216.
Edgecote, 183.
Ensham, 161.
Enstone, 41. 86. 142.
Ewelme, 20.
Great Handborough, 41.
Great Rollwright, 41. 43.
Hook Norton, 41.
Horley, 58. 201. 202.
Hornton, 202.
Kiddington, 29.
Marston, 148.
Minster Lovel, 91. 148.
Oxford, St. Frideswyde's, 253. 255. 274. 281.
——— Merton college, 73.
——— New college, 84.
——— Trinity college, 89. 90.
Sandford, 62. 88. 89.
Standlake, 92.
Stanton Harcourt, 12. 37. 120.
Warborough, 17.
Woodperry, 229.

### RUTLAND.

Lyddington, 129.
Tynwell, 21.
Whitwell, 160.

## SHROPSHIRE.

Alderbury, 21.
Battlefield, 59. 68. 182.
Buildwas abbey, 250. 251. 262. 265.
Burford, 150.
Ellesmere, 95.
Haughmond abbey, 263.
Ludlow, 47.
St. Kenelm's, 180.
Shrewsbury, 227. 238. 251. 254.
Stanton Lacey, 8.

## SOMERSETSHIRE.

Banwell, 41.
Bristol cathedral, 263.
——— Mayor's chapel, 92.
——— St. Mary's, Redcliffe, 106.
Chedzoy, 156.
Clapton-in-Gordano, 15. 83.
Glastonbury abbey, 10. 13. 241. 258. 300.
Kew Stoke, 152. 153.
Kingsbury Episcopi, 157.
Kingston Seymour, 16.
Langport, 157.
Long Sutton, 34. 40. 161.
Lullington, 19.
Montacute, 129.
Mutchingley abbey, 12. 58.
Nettlecomb, 89.
Othery, 58.
Portbury, 27. 30.
Portishead, 15. 27.
Tickenham, 30.
Trull, 32. 34.
Wellington, 277.
Wells cathedral, 139. 264.
——— Vicar's close, 145.
Weston-in-Gordano, 15. 27.
Wraxhall, 20.
Yeovil, 27. 73.

## STAFFORDSHIRE.

Clifton Campville, 179.
Fairwell, 155.
Sandwell priory, 27.
Tamworth, 191.
Wolverhampton, 70.

## SUFFOLK.

Bury St. Edmunds abbey, 33. 38. 189. 269.
Chevington, 161.
Dunwich, 69.
Elmham, 201.
Freslingham, 16.
Freshlingfield, 145.
Hawstead, 75.
Little Saxham, 23.
Long Melford, 68. 104. 105.
Lowestoft, St. Margaret's, 182. 192.
Mettingham, 181.
Pakefield, 192.
Stoke-by-Nayland, 18.
Sudbury, 21.
Ufford, 21.

## SURREY.

Charlewood, 199.
Chertsey abbey, 229.
Compton, 35.
Croydon, 71. 200.
Longfield, 75.
Southwark, St. Margaret's, 123.
Waverly abbey, 151.
Worth, 8.

## SUSSEX.

Arundel, 77. 138. 143. 179.
Battle, 200.
Brighton, St. Nicholas, 232.
Broadwater, 91.
Chichester cathedral, 65. 66. 281.
Climping, 161.
Preston, 202. 203.
Sele priory, (Beeding), 300. 301.
Steyning, 238.
Tarring Neville, 143.

## WARWICKSHIRE.

Astley, 23. 275.
Beaudesert, 12. 37.
Bilton, 120. 130. 142. 221.
Brailes, 27. 162. 182.
Brinklow, 41. 231.
Brownsover, 142.
Chesterton, 142.
Church Lawford, 28. 142.
Clifton-on-Dunsmore, 23.
Coleshill, 233.
Combe abbey, 231. 263. 265. 267.
Compton Wyniate, 145.
Coventry, Holy Trinity, 34. 71. 203.
——— St. Michael's, 47. 71. 106.
——— monastery, 57.
——— St. Mary's hall, 61.
——— St. John's, 169. 178.
——— White Friars, 246. 251.
Cubbington, 119.
Dunchurch, 61.
Hampton-in-Arden, 150.
Ilmington, 12.
Kenilworth, 129.
——— priory, 246. 264. 272.

Knowle, 275.
Lapworth, 83. 154.
Long Compton, 27.
Long Itchington, 37. 119. 219.
Mancetter, 220.
Maxstoke priory, 246.
Merivale abbey, 220. 248. 253. 254. 284. 300. 303.
Monks Kirby, 20. 23.
Newbold-on-Avon, 41.
Newnham Regis, 94.
Polesworth nunnery, 238. 295.
Prior's Hardwick, 92.
Rowington, 12. 34. 92.
Rugby, 21. 228.
Shotswell, 31. 37. 143.
Solihull, 139.
Stoneleigh abbey, 245. 246. 259. 262. 267. 271.
Stratford-upon-Avon, 92. 137. 193. 202. 203.
Stretton-on-Dunsmore, 94.
Tysoe, 32.
Warmington, 132. 143. 144.
Warwick, 28. 157. 202. 222. 281. 283.
Whichford, 27.
Withybrook, 20. 123.
Wormleighton, 41.
Wolfhampcote, 37.
Wootton-Wawen, 8.
Wroxhall priory, 262. 267.

### WILTSHIRE.

Boyton, 179.
Bradford-on-Avon, 8. 14.
Combe Bissett, 20.
Dilleridge, 201.
Great Bedwyn, 232.
Salisbury cathedral, 97. 139. 257. 263. 266.
Sherborne abbey, 250.
Stanton Fitzwarren, 19.

### WORCESTERSHIRE.

Alvechurch, 93.
Bengeworth, 142.
Churchhill, 161.
Cookhill, 68.
Crowle, 76.
Evesham, 76.
——— abbey, 249.
——— St. Lawrence, 44.
Great Malvern, 222. 232. 282.
Little Malvern, 43. 282.
Pershore abbey, 238.
Tenbury, 182.

Worcester cathedral, 48. 85. 88. 139. 187. 251. 252. 253. 255. 270. 271.
Witton, 232.

### YORKSHIRE.

Bedale, 140.
Beverley, St. Mary's, 19. 46. 289.
Bolton, 142.
Bridlington priory, 63. 160. 237. 238. 246. 247. 252. 255. 259. 260. 264. 270. 272.
Conisborough, 201.
Fountains abbey, 155. 251. 252. 253. 266. 268 274.
Howden, 264.
Kirkstall abbey, 251. 252. 259. 266. 275. 307. 309.
Patrington, 120. 122.
Ripon cathedral, 8. 9. 191. 239.
——— Maison Dieu, 142.
Roche abbey, 247. 275.
Salley abbey, 95.
Tanfield, 125.
Whitby abbey. 275.
York cathedral, 57. 84. 264.
——— St. Mary's abbey, 253. 262. 268. 281.
——— All Saints', 146.
——— St. Michael le Belfry, 146.
——— St. Sampson, 162.

### IRELAND.

Innisfallen abbey, Lakes of Killarney, 240. 241.
Newry abbey, County Down, 269.

### SCOTLAND.

KIRKCUDBRIGHTSHIRE—
Dundrennan abbey, 284.

### WALES.

ANGLESEA—
Llandegfan, 144.
Llanelian, 42.
Penmon priory, 253. 255.

BRECKNOCKSHIRE—
Crickhowel, 42.
Patricio, 42. 140. 178.

CAERNARVONSHIRE—
Bangor cathedral, 75.

## MONASTERIES REFERRED TO.

DENBIGHSHIRE—
Henllan, 21.
Llanwryst, 41.

GLAMORGANSHIRE—
Llangyfelach, 21.
Neath abbey, 230.

MERIONETHSHIRE—
Bettys Gwerful Goch, 42.
Llanderfell, 62.

MONMOUTHSHIRE—
Abergavenny, 62.
Caldicott, 15, 16.
Lanthony abbey, 250. 252. 263. 265. 273.
Tintern abbey, 250. 258. 261. 263. 265. 273.

PEMBROKESHIRE—
Gumfreston, 27.
St. David's cathedral, 141.
Tenby, 142.
Upton Castle, chapel of, 142.

### FOREIGN CHURCHES.

Amiens cathedral, 81.
Blanche Mortain abbey, 243.
Clairvaux, 187.
Neocæsarea, 6.
Oischot, 228.
St. Chapelle, Paris, 228.
St. Denis, Paris, 166.
Tyre, 7.
Vienne, St. Laurence, 164.

Sepulchral Monument, Westminster Abbey.
14th Century.

# INDEX

## OF NAMES OF PERSONS.

## VOL. II.

AETHELWOLD, 24.
Agnes, Dame, 170.
Alcuinus, Albinus Flaccus, 102.
Alexander Severus, 56.
Alfrida, 166.
Alild, King of Munster, 240.
Amalarius, Archbishop, 102.
Ambrose, St., 100.
Anneys, Dame, 171.
Apollonius, 56.
Arthur, Prince, 139.
Astley, John de, 81.
Audley, Bishop, 139.
Augustine, St., (of Hippo), 55. 98. 133.
———— (of Britain), 5. 8. 151. 216.

BASIL, Bishop of Cæsarea, 3.
Becket, Archbishop, 203. 213.
Becon, 175. 294.
Beaufort, Cardinal, 139.
Bedyll, Thomas, 128.
Bede, Venerable, 24. 93. 133. 197.
Bedewynde, Johanni de, 127.
Bega, St., 24.
Beke, Matthew, 40.
Benedict, St., 235.
Bernard, St., 235.
Berkeley, Lord, 151.
Berengarius, 80.
Bertramus, 80.
Birinus, St., 216.
Biscopius, Benedict, 9. 10. 197. 214.
Bleys, Bishop, 87.
Blois, William de, Bishop, 187.
Bloomfield, 105. 170.
Bolter, 29.
Boltoner, William, 72.
Bonner, Bishop, 13.
Boyle, Sir Philip, 81.
Brendan, St., 240.
Briton, 238.
Bridges, 196.
Brown, 211.
Bucks, Eleanor, Duchess of, 55.
Bruges, William, 39. 55.
Bubwith, Bishop, 139.
Buck, S. and N., 238.

CANYNE, MAISTER, 106.
Cantilupe, Walter de, Bishop, 49. 87.
Carpocrates, 55.
Charles, King of the Franks, 166.
Charlemagne, 56.
Chrysostom, St., 8. 77.
Claude de la Croix, 96. 98. 186.
Clement IV., Pope, 126.
Compton, Robert de, 121.
Constantine, 99.
Cornwall, Edmund, 150.
Courtney, Archbishop, 172.
Cranmer, Archbishop, 114.
Crumwell, 129.
Cyril, St., 93. 100.
Cuthbert, St., 87. 159.

DALTON, RICHARD DE, 163.
Darcy, Sir Philip, 64.
Davies, of Kidwelly, 13. 238.
Dele, Willelmo de, 127.
De Vert, 95. 187.
Dervel Gadarn, 8. 62.
Dineley, 43.
Drope, Robert, 40.
Du Cange, 50. 53.
Dugdale, Sir William, 81.
Durand, 187.
Durandus, 26. 35. 102.
Durantus, 11.
Dunkin, 29.
Dunstan, St., 278. 279. 280.

EGBERT, Archbishop, 24. 49. 77.
Edgar, 24.
Edgin, St., Bishop, 76.
Edington, Bishop, 4.
Edrygge, Agnes Dame, 170.
Edward the Confessor, King, 278.
Edward II., King, 163.
Edward III., King, 214.
Edmund, Archbishop, 17.
Egelricus, Abbot, 241.
Ellacombe, 29.
Ellis, Sir Henry, 238.
Elsee, 29.
Emma, Lady, 172.

## INDEX OF NAMES OF PERSONS. 323

Epiphanius, 55.
Ethelbald, St., 278.
Eugenius IV., Pope, 172.
Eusebius, 7. 100.

FAIRHOLT, 48.
Fane, John, 39.
Finian, St., 240.
Fisher, T., 204.
Forest, Friar, 62.
Fox, 6. 136.

GELASIUS, 5. 101.
Gervase, 39.
Gildas, 8.
Gibbs, John, 185.
Gilpin, George, 113.
Giffard, Archbishop of York, 128.
——— Bishop of Winchester, 151.
Glanville, Bishop, 281.
Godric, St., 159.
Googe, Barnabe, 112.
Gondibour, Prior of Carlisle, 213.
Godiva, Countess of Mercia, 57.
Gregory the Great, 5. 56. 101. 133. 151.
Gregory, Thaumatergus, 6.
Grimlaic, 165. 166.
Grimo, Abbot, 166.

HARDING, DR., 83. 293. 294.
Harvey, Elizabeth, Abbess, 295.
Hastead, 194. 195.
Helen, St., 99.
Helyot, 169.
Henry II., King, 157.
——— III., King, 200.
——— IV., King, 203. 204.
——— VII., King, 204.
Hidelburgis, 167.
Hilda, St., 24.
Hope, 29.
Hopkins, 48.
Husenbeth, F.C., D.D., 213.

INA, 10.
Ingham, Sir Oliver, 204.
Innocent III., Pope, 80. 81. 158.
Ireneus, 55.
Isabell, Countess of Warwick, 139.
Isidore, St., 102.
Ithamar, St., 281.

JEWELL, Bishop, 83. 98. 99. 132. 136. 293. 294.
Jewitt, 29.
Joan, Lady, 171.
Jocelin de Brakelond, 38. 159. 266.
John of Beverley, St., 278.

John, King, 85.
John de St. Omer, 71.
John XIII., Pope, 28.
Justin, Martyr, 2. 3.

KENELDEN, JOHANNI DE, 127.
Kip, 238.
Kirkham, Walter de, Bishop, 50.
Knyghton, Henry de, 172.

LACY, Bishop, 146. 168. 169.
Lanfranc, Archbishop, 24. 39. 77. 80.
Latimer, Bishop, 36. 70.
Leland, 194. 269.
Leo the Thracian, 124.
——— III., Pope, 56.
——— IV., Pope, 12. 56.
L'Estrange, 29.
Leofric, Earl of Mercia, 57.
Leonianus, Abbot, 164.
Littledale, 4.
Lisle, Joan, 39.
Lyndwood, 17. 147.
Lyra, Nicholas de, 269.

MABILLON, 5. 164. 167. 237.
Malemeyns, Rogero, 127.
Malmesbur, Thomæ de, 127.
Malory, Sir Thomas, 173.
Margaret, Countess of Salisbury, 139.
Martene, 12. 103. 187. 236.
Marlebugh, Thomas de, 76.
Martial, St., 67.
Martin, Roger, 104.
Matilda, 173.
Masse, Peter de, 81.
Maurus, Rabanus, 102.
Mellitus, 151.
Montacute, Lady Elizabeth, 296.
Moresme, Richard, 269.
Morgan, Octavius, Mr., 89.
Muratori, 5.

NAOGEOROUS, THOMAS, 112.
Neale, 4.
Norman, Margaret, 171.
North, 29.

ORIGEN, 132.
Orpheus, 56.
Osmund, Bishop of Salisbury, 6.
Oswald, King, 93.

PARIS, MATTHEW, 87. 133.
Parker, Abbot, 139.
Parkin, 105.
Pascasius, Radbertus, 79. 80.
Paulinus, Bishop of Tyre, 7. 281.

# INDEX OF NAMES OF PERSONS.

Peacock, 115.
Peckham, Archbishop, 51.
Place, William, 268.
Plato, 132.
Pole, Cardinal, 114.
Pope, Sir Thomas, 90.
Prudde, John, 222. 223.

RADEGUND, ST. 165.
Raine, 107.
Raven, Dr. Rev. 29.
Renaudot, 4.
Reynold, Walter, Archbishop, 124.
Richard, Prior of Hexham, 9. 197.
Richard I., King, 87.
Richard II., King, 214.
Richard, Bishop of Chichester, 168.
Richard, St., 281.
Roger, Bishop of Coventry, 169.
Rowland, D., Rev. 227.
Rumor, St., 278.

SALISBURY, Earl of, 54.
Salmon, John, Bishop of Norwich, 190.
Sampson, 159.
Scott, Ann, 185.
Scott, Elizabeth, Dame, 170.
Scotus, 80.
Scroop, Thomas, 172.
Scroop, Henry, third Lord, 174.
Scribum, Nicholas de, 162.
Sherbrook, Henry, 169.
Sigibert, 166.
Smith, Chamberlayne, 72.
Stafford, John, 26.
Starkey, Richard, 40.
Staunton, William, 28.
Staunton, Adam de, 160.
Stephen, Archbishop, 147.
Stigand, Archbishop, 38.
Sugar, Dean, 139.
Swyndurby, William de, 173.

TERTULLIAN, 3.

Theobald, Archbishop, 278.
Theophilus, Emperor, 56.
Theudcrius, 164.
Townsend, Sir Roger, 106.
———— Eleanore, 106.
Trollope, Bishop Suffragan of Nottingham, 48.
Tylle, Richard, 55.
Tyldisley, Thurstan, 40.
Tyssen, 29.

WADDING, 126.
Wainflete, Bishop, 139.
Walton, William, 40.
Walter of Colchester, 199.
Wallingford, Abbot, 277.
Warin, 159.
Wendreda, St., 278.
Wenefred, St., 278.
Werburgh, St., 278.
Westminster, Matthew of, 87.
Whiston, 7.
Whalley, Peter, 196.
Whittington, Richard, 269.
Wilfred, 9. 214.
Wilfrid, St., Archbishop, 197.
William of Malmesbury, 57.
Wilsius, St., 78.
William I., King, 87.
Wilkins, 103. 126. 127.
William of Wyckham, Bishop, 139.
Willis, Brown, 238.
Willis, Professor, 244.
Winchelsey, Archbishop, 53.
Winchelsey, Thomas, Dr., 269.
Windsor, Thomas, 105.
Wolsharte, Robert, 201.
Wolsey, Cardinal, 300.
Wodeheye, Radulpho de, 127.
Wren, Bishop, 19.
Wright, 48.
Wycliffe, John, 172.

ZOSIMUS, 11.

Monument, Hillmorton Church, Warwickshire.

PRINTED BY A. J. LAWRENCE, RUGBY.

www.ingramcontent.com/pod-product-compliance
Lightning Source LLC
Chambersburg PA
CBHW021209230426
43667CB00006B/628